VOLUME ONE HUNDRED AND EIGHTY FIVE

PROGRESS IN MOLECULAR BIOLOGY AND TRANSLATIONAL SCIENCE

Advances in Aggregation Induced Emission Materials in Biosensing and Imaging for Biomedical Applications - Part B

VOLUME ONE HUNDRED AND EIGHTY FIVE

PROGRESS IN MOLECULAR BIOLOGY AND TRANSLATIONAL SCIENCE

Advances in Aggregation Induced Emission Materials in Biosensing and Imaging for Biomedical Applications - Part B

Edited by

RAJESH S. BHOSALE
Department of Chemistry, School of Science, Indrashil University, Rajpur, Mehsana, Gujarat, India

VIJAI SINGH
Department of Biosciences, School of Science, Indrashil University, Rajpur, Mehsana, Gujarat, India

ELSEVIER

ACADEMIC PRESS
An imprint of Elsevier

Academic Press is an imprint of Elsevier
50 Hampshire Street, 5th Floor, Cambridge, MA 02139, United States
525 B Street, Suite 1650, San Diego, CA 92101, United States
The Boulevard, Langford Lane, Kidlington, Oxford OX5 1GB, United Kingdom
125 London Wall, London EC2Y 5AS, United Kingdom

First edition 2021

ISBN: 978-0-323-99604-4
ISSN: 1877-1173

For information on all Academic Press publications
visit our website at https://www.elsevier.com/books-and-journals

Publisher: Zoe Kruze
Acquisitions Editor: Ashlie M. Jackman
Developmental Editor: Jhon Michael Peñano
Production Project Manager: James Selvam
Cover Designer: Matthew Limbert

Typeset by STRAIVE, India

Contents

Contributors

Laxmi Raman Adil
Department of Chemistry, Indian Institute of Technology Guwahati, Guwahati, Assam, India

Debasish Barman
Department of Chemistry, Indian Institute of Technology Guwahati, Guwahati, Assam, India

Rajesh S. Bhosale
Department of Chemistry, Indrashil University, Rajpur, Mesana, Gujarat, India

Sidhanath Vishwanath Bhosale
Polymers and Functional Materials Division, CSIR-Indian Institute of Chemical Technology, Hyderabad, Telangana; Academy of Scientific and Innovative Research (AcSIR), Ghaziabad, Uttar Pradesh, India

Madan R. Biradar
Polymers and Functional Materials Division, CSIR-Indian Institute of Chemical Technology, Hyderabad, Telangana; Academy of Scientific and Innovative Research (AcSIR), Ghaziabad, Uttar Pradesh, India

Shailesh S. Birajdar
Polymers and Functional Materials Division, CSIR-Indian Institute of Chemical Technology, Hyderabad, Telangana; Academy of Scientific and Innovative Research (AcSIR), Ghaziabad, Uttar Pradesh, India

Sudesna Chakravarty
Department of Biochemistry and Molecular Biology, University of Nebraska Medical Center (UNMC), Omaha, Nebraska, United States

Rajdikshit Gogoi
Department of Chemistry, Indian Institute of Technology Guwahati, Guwahati, Assam, India

Parameswar Krishnan Iyer
Department of Chemistry; Centre for Nanotechnology; School of Health Sciences and Technology, Indian Institute of Technology Guwahati, Guwahati, Assam, India

Mst Nasima Khatun
Department of Chemistry, Indian Institute of Technology Guwahati, Guwahati, Assam, India

Jotiram N. Malegaonkar
Polymers and Functional Materials Division, CSIR-Indian Institute of Chemical Technology, Hyderabad, Telangana; Academy of Scientific and Innovative Research (AcSIR), Ghaziabad, Uttar Pradesh, India

Tahir Mehmood
Department of Chemistry, School of Sciences, Indrashil University, Rajpur, Gujarat, India

Sudip Mukherjee
Department of Bioengineering, Rice University, Houston, TX, United States

Kavita Narang
Department of Chemistry, Indian Institute of Technology Guwahati, Guwahati, Assam, India

Retwik Parui
Department of Chemistry, Indian Institute of Technology Guwahati, Guwahati, Assam, India

Karma Patel
School of Chemical Sciences, Central University of Gujarat, Gandhinagar, India

Panchami Prabhakaran
School of Chemical Sciences, Central University of Gujarat, Gandhinagar, India

J. Prakasha Reddy
Department of Chemistry, School of Sciences, Indrashil University, Rajpur, Gujarat, India

Sayan Roy Chowdhury
Department of Chemistry, Virginia Commonwealth University, Richmond, VA, United States

Vishal Saini
Department of Chemistry, Indian Institute of Technology Roorkee, Roorkee, Uttarakhand, India

Syed Kabir Hussain Shah
School of Chemical Sciences, Central University of Gujarat, Gandhinagar, India

Arvin Sain Tanwar
Department of Chemistry, Indian Institute of Technology Guwahati, Guwahati, Assam, India

V. Venkatesh
Department of Chemistry, Indian Institute of Technology Roorkee, Roorkee, Uttarakhand, India

Sopan M. Wagalgave
Polymers and Functional Materials Division, CSIR-Indian Institute of Chemical Technology, Hyderabad, Telangana; Academy of Scientific and Innovative Research (AcSIR), Ghaziabad, Uttar Pradesh, India

Nehal Zehra
Department of Chemistry, Indian Institute of Technology Guwahati, Guwahati, Assam, India

Preface

The understanding of cellular-level biological events is of great importance for resolving the closely associated health issues of living organisms. Fluorescence-based monitoring and diagnosis platforms furnish extensive insight into the molecular mechanisms of cellular processes and precise guidance of therapeutics. Traditional/conventional fluorophores exhibit high emission in the solution state and lose their emission in aggregation and solid forms. This loss of emission in aggregation and solid forms limits conventional fluorophore utilization in monitoring and diagnosis of biological processes. In the past decade, aggregation induced emission (AIE) active small molecules and polymers have played a vital role over traditional fluorophores in the understanding of various cellular-level biological processes. The development of AIE-based systems allows worldwide healthcare and biological researchers to monitor the molecular level pathogenesis of diseases. Owing to the fundamental and practical importance of AIE-based materials, researchers worldwide, including material chemists, photophysicists, and biologists, have successfully utilized AIE materials in aggregate/solid states for various advanced applications, such as chemo/biosensing, imaging, biomedical, organic electronics, and many more.

This volume consists of eight chapters, which represent current efforts addressing advances in AIE materials and their insightful biomedical applications. Chapter 1 covers the last two decades of important literature of AIE materials for tissue-imaging applications. Chapter 2 reveals the application of AIE materials for cancer cell detection, bioimaging, and theranostics. Chapter 3 deals with recent advancements in AIE material for photodynamic therapy. Chapter 4 mainly discusses advanced photophysical properties of AIE materials and their biomedical applications. Chapter 5 reveals the protein fibril imaging applications of AIE active materials. Chapter 6 gives special emphasis to AIE active polymers for bioimaging applications. Chapter 7 mainly discusses AIE-MOF materials and their biological applications. Chapter 8 is specifically dedicated to patented AIE materials and their biomedical applications.

This volume will help future investigators, researchers, students, and stakeholders to perform their research with greater ease. It will also serve as an excellent basis from which scientific knowledge can grow and widen in the field of AIE and its applications in different areas.

Rajesh S. Bhosale
Vijai Singh

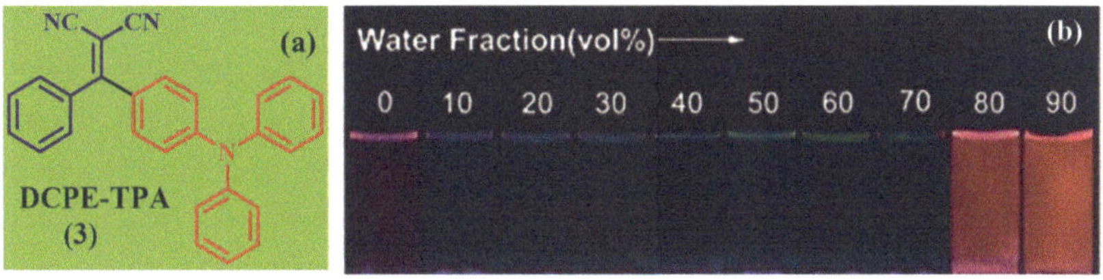

Fig. 3 (A) Molecular design of probe **3**; (B) necked eye image of THF and water mixtures at various water percentage (*in* vol%). *Reproduced with permission from Tian H, Li D, Tang X, Zhang Y, Yang Z, Qian J, Dong YQ, Han M. Efficient red luminogen with aggregation-induced emission for: in vivo three-photon brain vascular imaging.* Mater Chem Front. *2020;4:1634–1642. Copyright @ 2020, Royal Chemical Society.*

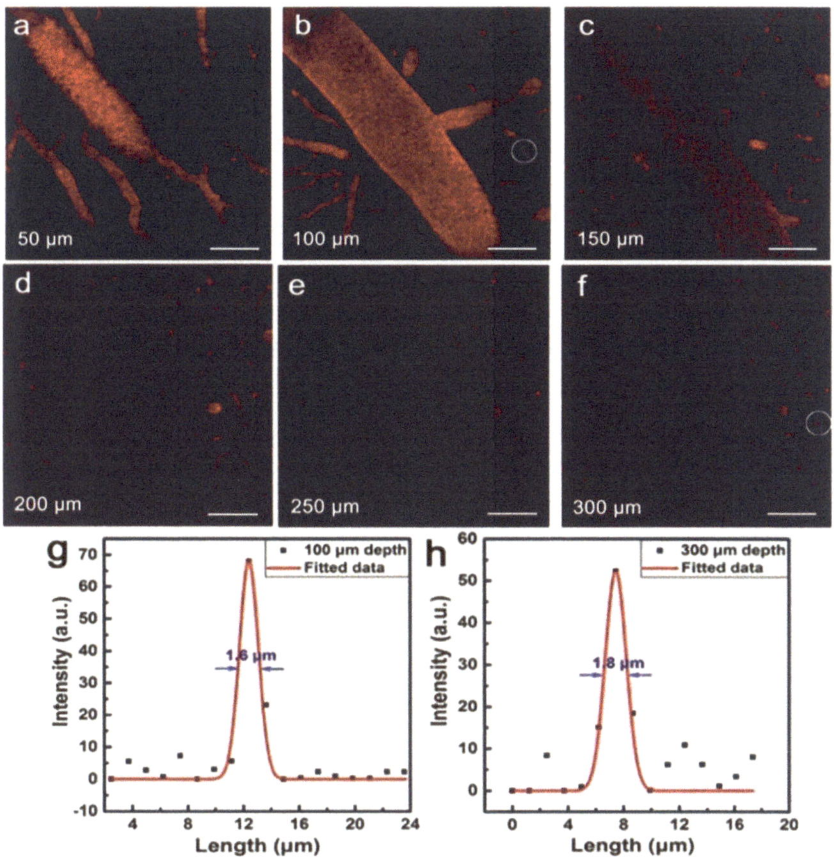

Fig. 4 (A–F) In vivo 3-PM images of brain tissue of a mice at different depth; (G) intensity profile study along the line in (B); showing resolution of 1.6 μm at a depth of 100 μm; (H) intensity profile study along the line in (F); showing resolution of 1.8 μm at a depth of 300 μm. *Reproduced with permission from Tian H, Li D, Tang X, Zhang Y, Yang Z, Qian J, Dong YQ, Han M. Efficient red luminogen with aggregation-induced emission for: in vivo three-photon brain vascular imaging.* Mater Chem Front. *2020;4:1634–1642. Copyright @ 2020, Royal Chemical Society.*

CHAPTER ONE

Aggregation induced emission materials for tissue imaging

Madan R. Biradar[a,b], Rajesh S. Bhosale[c,*], and Sidhanath Vishwanath Bhosale[a,b,*]

[a]Polymers and Functional Materials Division, CSIR-Indian Institute of Chemical Technology, Hyderabad, Telangana, India
[b]Academy of Scientific and Innovative Research (AcSIR), Ghaziabad, Uttar Pradesh, India
[c]Department of Chemistry, Indrashil University, Rajpur, Mesana, Gujarat, India
*Corresponding authors: e-mail addresses: rajeshbhosale24@gmail.com; sidhanath.bhosale@gmail.com; bhosale@iict.res.in

Contents

Abstract

Aggregation Induced Emission (AIE) has ample opportunities in the sensing and biomedical field. Recently, AIE molecular architecture, AIE nanoparticles (NPs) and AIE conjugated polymer (CP) probes have shown outstanding performance in bioimaging applications. In this chapter, we are summarizing the updated advancement in AIE based molecular, NP and CP probes for tissue imaging. We are focusing in vivo and in vitro tissue imaging with in-depth morphological and molecular information using microscopic techniques.

Progress in Molecular Biology and Translational Science, Volume 185
ISSN 1877-1173
https://doi.org/10.1016/bs.pmbts.2021.07.005

Also, we have summarized various infrared windows for fluorescence microscopic technique to achieve deep penetration and high resolution of tissue images. In addition, future difficulties and challenges of AIEgens in tissue imaging are shortly discussed.

1. Introduction

In the previous few years, fluorescent imaging innovation has made incredible achievements in the biomedical area.[1] The utilization of fluorogen is exceptionally limited by the aggregation caused quenching (ACQ) effect.[2] There are numerous organic luminophore shows different light-emitting behaviors in low and high concentration solutions; at high concentration luminescence is often quenched, known as aggregation-caused quenching (ACQ).[3] This ACQ impactis created because of regular luminophores are involved in planar aromatic rings. To avoid the ACQ, the molecular architecture has developed by increasing the extention of π-conjugation by fusing more and more aromatic rings together.[4] The molecules are non-emissive in low concentration solution but became highly luminescent when they were aggregated in highly concentrated solutions. The light emission was produced by aggregate formation known as aggregation induced emission (AIE).[5] The remarkable optical properties of AIEgens make them ideal material for various imaging applications.[6]

Tissue is the bunch of cells having a similar structure and function as that of a single cell. Tissue images play an important role in analyzing tumors as well as the drug metabolism process.[7,8] There are various bioimaging techniques developed such as computed tomography (CP), magnetic resonance imaging (MRI), ultrasound (US), positron emission tomography (PET), mass spectrometry (MS) and fluorescence microscopy.[9–12]

Tissue imaging by MS is useful for studying metabolism, proteins and lipids in tissues.[13] Although fluorescence imaging also rapidly developing because of its silent features such as high sensitivity, rapid response, absence of ionizing radiation and high resolution of images. Most of the structures and components in biological samples don't emit fluorescence, when AIE active probe binds to an analyte it shows fluorescence under microscopy.[14] Hence with the microscopic techniques, there is a need to develop AIE active molecular probe, AIE active NP probe and AIE active CP probe.[15–18]

In the literature the role of different AIEgens such as molecular probe containing proteins, NP probe containing quantum dots, carbon dots and other novel nanoparticles are designed for biomedical applications.[19,20]

In this section, we summarize the latest progress in AIE probes, CP probes and NP probes as luminescent specialists for deep tissue imaging including near-infrared (NIR) fluorescence, chemiluminescent, and photon fluorescence microscopic images.[21,22]

2. Mass spectrometric techniques for tissue imaging

The most commonly used MS techniques are matrix-assisted laser desorption/ionization mass spectroscopy (MALDI-MS), Laser Ablation Inductively Coupled Plasma Mass Spectrometry (LAICPMS) and secondary ion mass spectrometry (SIM).[23,24] LAICPMS technique is utilized for tissue imaging containing biomolecules and metals. MALDI-MS is useful for analyzing and distributing relative compounds within the tissue including lipids and proteins. MALDI is used to image whole-body tissue to observe the mapping of drug distribution and its metabolites at different dosages.[25]

Recently, Brunelle et al. reported the role of MALDI in a very rare disease known as pulmonary arterial hypertension. In this study, they had used lung tissue for MS imaging.[26] With these advantages this technique has some drawbacks such as sample is analyzed under vacuum conditions. This is also an invasive technique where the sample is not recovered.[27]

3. Fluorescence spectroscopic techniques for tissue imaging

To overcome the disadvantages of MS imaging technique, fluorescence spectroscopic technique is widely used for its advances such as useful for in-vivo imaging, shows abundant signals and fast reposing, ionization radiation-free and have high spatial resolutions.[28,29] Along with these advantages, it has some drawbacks such as poor photon penetration depth in tissue.[30] The common fluorescence spectroscopic techniques are confocal fluorescence microscopy, single-photon fluorescence microscopy (1-PM), two-photon fluorescence microscopy (2-PM), and three-photon fluorescence microscopy (3-PM). These are the most widely adopted method for tissue imaging due to their rapid three dimensional imaging with high resolution.[31] The fluorescence intensity depends on the interaction between photon and tissue, causes absorption, reflection, scattering and autofluorescence, which is related to the penetration depth of tissue (Fig. 1A).[32] The excitation light for a biological sample is performed under different electromagnetic regions, such as the visible region (400–700 nm),

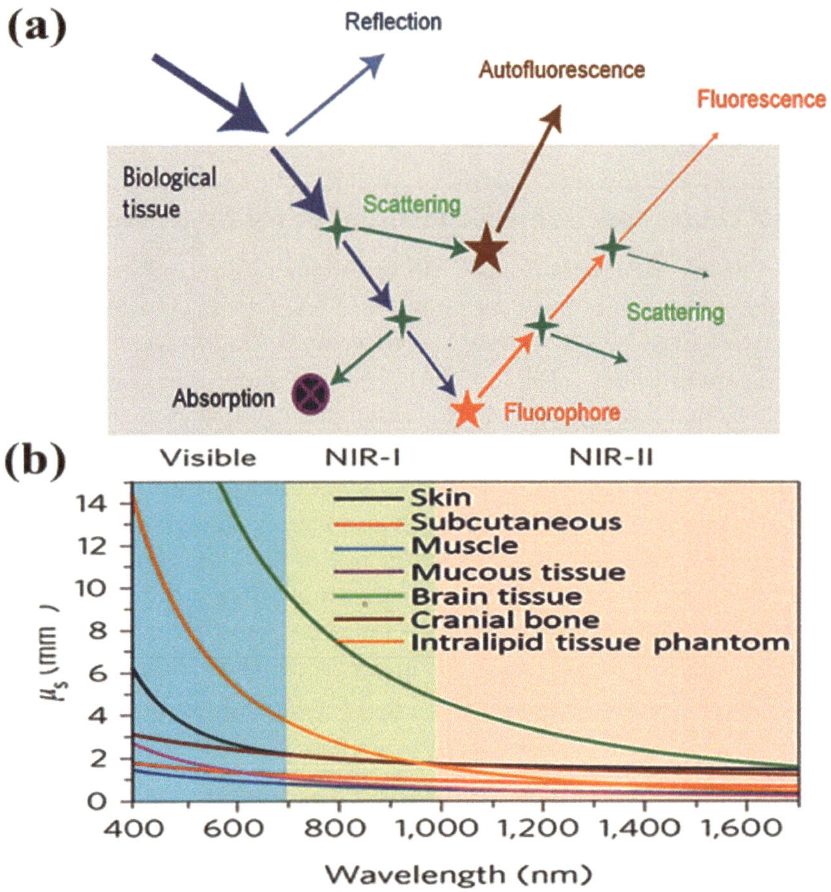

Fig. 1 (A) Diagram showing the interaction between photon and tissue; (B) scattering pattern of different biological tissue and intralipid solution in the wavelength range 400–1700 nm. *Reproduced with permission from Li C, Chen G, Zhang Y, Wu F, Wang Q. Advanced fluorescence imaging technology in the near-infrared-ii window for biomedical applications.* J Am Chem Soc. *2020;142:14789–14804. Copyright @ 2020, American Chemical Society.*

NIR-1 region (700–1000) and NIR-2 region (1000–1800 nm).[32,33] The tissue penetration power of visible light is near about 1 mm which causes autofluorescence of tissue. NIR-1 has a light penetration depth is near about 1.0–3.5 mm, which is suitable for fluorescence tissue imaging. Still, it has some issues such as auto fluorescence and poor imaging depth.[34] NIR-II has tissue penetration depth is about 5–20 mm, which is suitable for fluorescence images with very low background noise.[35] In the NIR region there is very low autofluorescence and decreased scattering of light as shown in Fig. 1B. It indicates the importance of excitation of light for deep tissue imaging.

4. AIE molecular probes for deep tissue imaging

4.1 Detection and imaging of singlet oxygen in tissue during photodynamic therapy using 2-PM

Tan et al. in 2016 reported 2-PM probe containing 9-methylanthracene as an electron-accepting group for photoinduced electron transfer (PET) process, naphthalimide as a fluorophore and triphenylphosphonium salt as mitochondria targeting group. In the PET process probe MNAH **(1)** converts to strong fluorescent MNAO **(2)** by forming endoperoxide ring in methylanthracene group. (Fig. 2A) For the detecting and imaging of singlet oxygen ($^1O_2$) in liver tissue, performed by irradiating probe **(1)** with Ce6 at 635 nm for 6 min, exhibits strong fluorescence for the tissue imaging. The 2-PM technique is used for tissue imaging with penetration depth from 57 to 156 μm. as illustrated in Fig. 2B, this is an excellent probe for detecting and imaging liver tissue.[36]

The mechanochromism and secondary aggregation enhanced emission properties are useful for tissue imaging. Chi et al. used these concepts for long-term tumor tracking. They have synthesized TCPyP as an AIE active probe with photoluminescence studies in THF/water mixture. This luminescent molecule mainly injected neoplastic tissues giving high emission intensity. It is used for in-vivo imaging for nude mice using 2-PM technique.[37]

4.2 In vivo brain tissue imaging using 3-PM

Han et al. reported red luminogenic molecular probe DCPE-TPA (**3**) containing donor and acceptor group such as triphenylamine (TPA) and cyano with a quantum yield of 29.3% (Fig. 3A). This probe is used for in vivo brain tissue imaging with 3-PM technique. For the photoluminescence study, THF/water mixture was used with different fractions. In this system, different fraction water is used which influences the increased polarity of the solvent mixture. As shown in Fig. 3B, due to an increase in polarity of mixture solvatochromic effect was observed.[38]

For the brain tissue images, 3-PM technique is used to take high-resolution images. The excitation of light takes place at the focal point of the fluorophore. At the depth of 100 μm, shows the width of blood vessels

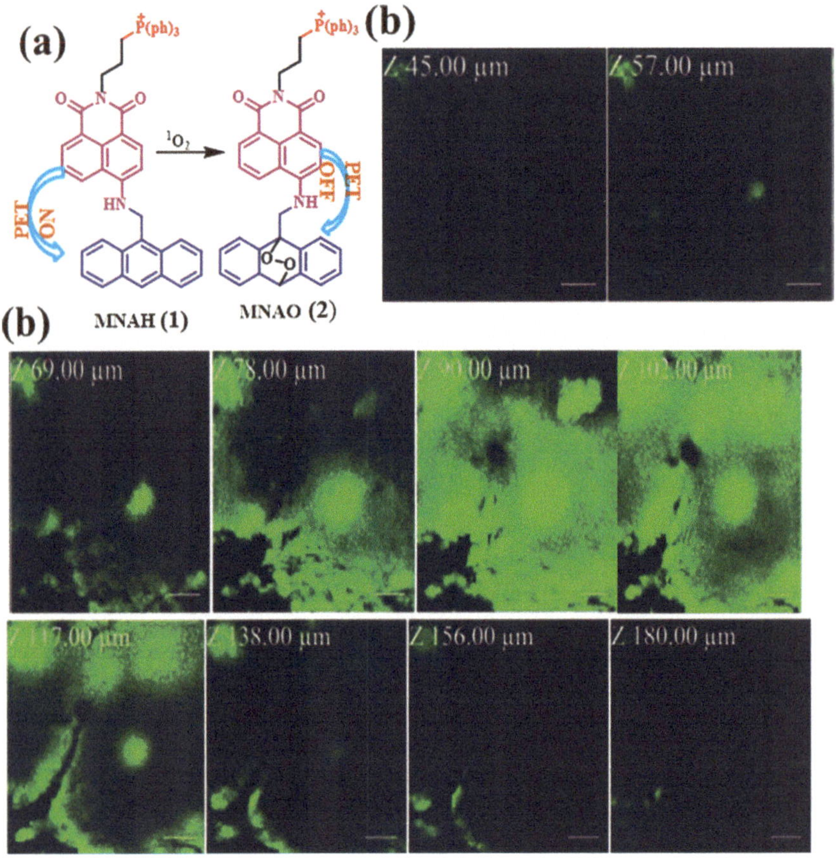

Fig. 2 (A) Molecular structure of probe **1** and **2**; (B) 2-PM images of rat liver frozen slice. Liver slides stained with probe (**2**) and chlorin e6 for 1 h. *Reproduced with permission from Liu HW, Xu S, Wang P, Hu XX, Zhang J, Yuan L, Zhang XB, Tan W. An efficient two-photon fluorescent probe for monitoring mitochondrial singlet oxygen in tissues during photodynamic therapy.* Chem Commun. *2016;52:12330–12333. Copyright @ 2016, Royal Chemical Society.*

1.6 μm (Fig. 4B and G). As the signal decreases, the resolution of the image increases such as intensity profile study along with the depth of 300 μm, shows resolution of 1.8 μm (Fig. 4F and H). This study shows probe **(3)** is the best for in vivo tissue images by using high-resolution 3-PM technique.[38]

4.3 Barbituric acid–TPE as AIEgens for high-resolution tissue images of secondary growth tumor

In 2018, Ding et al. reported three AIE gens based on D-π-A architecture. They have used TPE as electron donating and AIE active group, barbituric acid as an electron-accepting group and different π-conjugated unit for bridging the donor and acceptor group (Fig. 5A). The AIE property is observed in probe (**4**) to (**6**) in THF/water mixture. Due to the TICT effect and solvent polarity, photoluminescence (PL) intensity decreases by adding water fraction in THF. As illustrated in Fig. 5B–C, as water fraction increases above 60%, there is a abrupt expansion in PL intensity because of the aggregates in probe (**4**) and (**5**). In the case of probe (**6**), DMSO/toluene solvent system was employed for AIE studies. (Fig. 5D). As the volume of toluene increases, probe (**6**) exhibits blue shift with enhanced emission, such AIE properties are useful for further studies.[39]

For the tissue imaging, in vivo experiment is performed using probe (**6**), which is injected through intravenous injection for identifying metastatic tumor and their boundaries to normal tissue with high resolutions. In vivo tumor tissue analysis has done by injecting probe (**5**) in mice for 24 h (Fig. 6A). In the ex-vivo fluorescence image suggests that the AIE active probe (**5**) accumulate into the lung tissues of mice. As shown in Fig. 6B–D, the hematoxylin (H) and eosin (E) analysis show that the probe (**5**) visualized tissues in tumors (Fig. 6E). These results in vivo suggest that the NIR active probe (**5**) is an excellent candidate for metastatic tumor lung tissue analysis.[39]

5. AIE nanoparticle probe for tissue imaging

AIE probes show extraordinary future in the development of fluorescent NP probe with great quantum yields attributable to the limitation of intermolecular vibrational and rotational motions in the arrangement of nanoaggregates. As of late, different plans have been created to fabricate AIE-encapsulated nanoparticles for tissue imaging applications.[40,41]

5.1 Near-infrared AIE nanodots for tissue imaging

The fluorescence imaging In vivo NIR-II have great advantages such as high penetration depth and advanced resolution images due to negligible light scattering and decreased autofluorescence of tissue. Qian et al. and coworkers developed nanoparticles from amphiphilic polymer BPN-BBTD for in vivo bladder tumor imaging.[42] Organic fluorophores based on nanoparticles have a lot of advantages over molecular probes such as low cytotoxicity, high photostability and divert modifications on the surface. Tang et al. reported TPE-TPA-DCM (**7**) based nanoparticles (Fig. 7A). They employed serum albumin (BSA) as a polymer matrix for NPs formation. The fluorogen-loaded BSA NPs is synthesized as THF solution of probe (**7**) and watery solution of

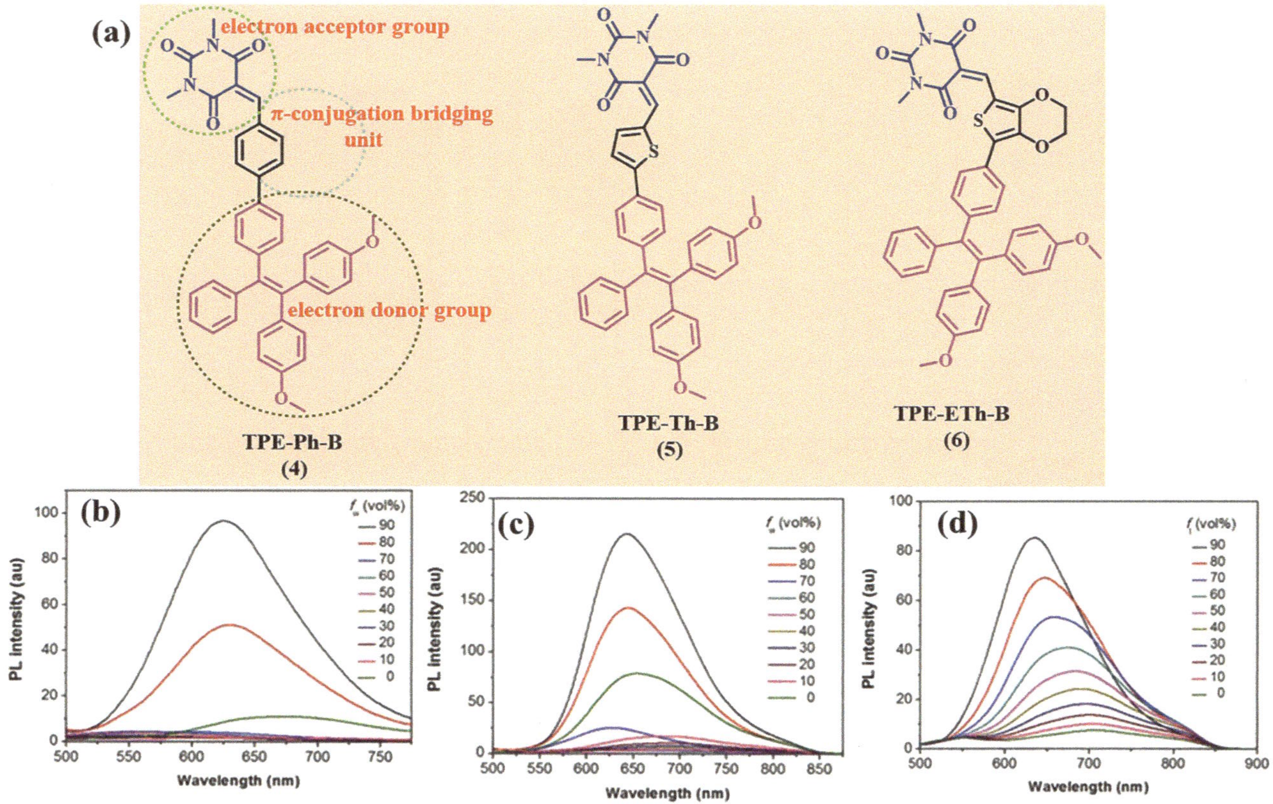

Fig. 5 (A) Molecular design of probe (**4**) to (**6**); (B–C) PL spectra of probe (**4**) to (**5**) in THF and water mixtures at different water percentage; (D) PL spectra of probe (**6**) in DMSO and toluene mixtures at different toluene percentage. *Adapted with permission from Gao H, Bao P, Dai S, Liu R, Ji S, Zeng S, Shen J, Liu Q, Ding D. Far-red/near-infrared emissive (1,3-dimethyl)barbituric acid-based aiegens for high-contrast detection of metastatic tumors in the lung.* Chem Asian J. *2019;14:871–876. Copyright @ 2018, Royal Chemical Society.*

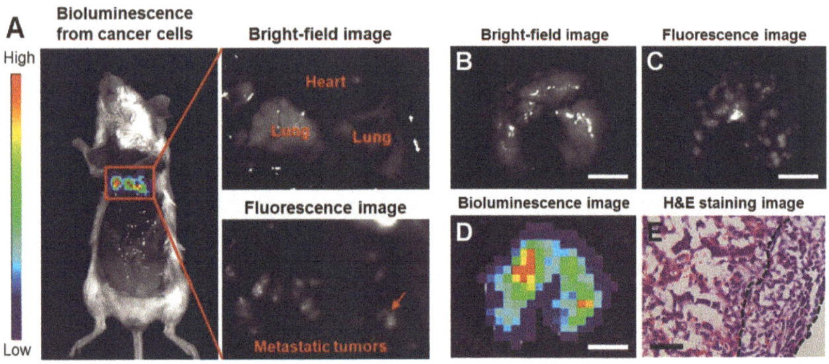

Fig. 6 (A) In vivo bioluminescence image, bright-field image and fluorescence image of pulmonary metastatic tumor tissue of mice; (B–D) Ex-Vivo bright filed image, fluorescence image and bioluminescence image; (E) H and E stained image. *Reproduced with permission from Gao H, Bao P, Dai S, Liu R, Ji S, Zeng S, Shen J, Liu Q, Ding D. Far-red/near-infrared emissive (1,3-dimethyl)barbituric acid-based aiegens for high-contrast detection of metastatic tumors in the lung.* Chem Asian J. *2019;14:871–876. Copyright @ 2018, Royal Chemical Society.*

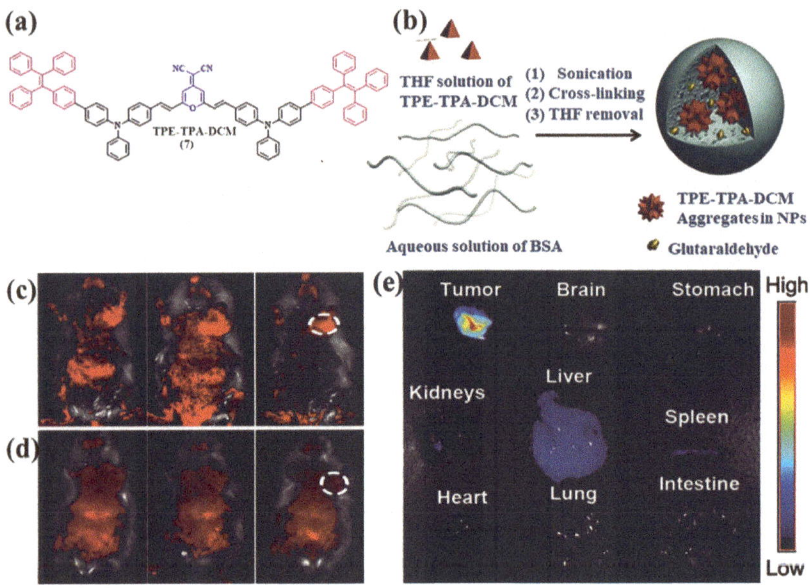

Fig. 7 (A) Molecular design of probe TPE-TPA-DCM (**7**); (B) Mechanism of NPs formation with probe (**7**); (C and D) in vivo fluorescence image of H_{22}-tumor bearing mouse after intravenous injection of fluorogen-loaded BSA NPs (C) and bare probe (**7**) NPs (D) at the same concentration of fluorogen. The white circles showed the tumor sites; (E) Ex-vivo fluorescence imaging on tumor tissue and body organs of mice treated with BSA NPs. *Reproduced with permission from Qin W, Ding D, Liu J, Yuan WZ, Hu Y, Liu B, Tang BZ. Biocompatible nanoparticles with aggregation-induced emission characteristics as far-red/near-infrared fluorescent bioprobes for in vitro and in vivo imaging applications.* Adv Funct Mater. *2012;22:771–779. Copyright @ 2012, Wiley-VCH.*

BSA, sonication using cross linker glutaraldehyde, after sonication fluorogenated BSA NPs formed immediately. The outer layer ionized carboxylic groups are useful for stabilizing NPs (Fig. 7B). PL spectra of probe (**7**) in THF and water mixtures, at various water volume percentage, which enabled sharp changing of the solvent polarity and the degree of solute aggregation. The probe (**7**) has an intense red fluorescence in THF solution, that's notably quenched when a modest amount of water was added.

For the in-vivo images, tumor bearing mice loaded with BSA NPs probe are used for fluorescence imaging. The BSA NPs light up the cancer tissue as shown in Fig. 7C. The strong fluorescence was seen in abdomen and liver regions at 3 h, fluorescence intensity completely disappeared after 28 h. Ex-vivo fluorescent images of tumor tissue and major organs like brain, stomach, kidney, liver, spleen, heart, intestine and lung had taken 28 h with the loading of BSA NPs probe (Fig. 7D and E).[43]

Tang group in 2020, published their work on chemiluminescent NIR active AIE dots for deep tissue imaging. They have synthesized TBL dots using electron donor and acceptor group such as triphenylamine and benzothiazole with F127 as the surfactant. TBL dots are distinguished between tumor tissues from normal by lighting up the high level of ROS in tumor environment, which shows the very bright application of the TBL dot probe for cancer diagnosis and surgery.[44]

5.2 Photon fluorescence microscopy for tissue imaging using AIE dots

In 2013, Liu et al. reported a new fluorescent organic material based AIE dot system. For the formation of aggregated luminescent nanoparticles they had used two organic fluorophores BTPETD (**8**) and TPETPAFN (**9**) as shown in Fig. 8A. The AIE dots (**10**) and (**11**) were synthesized utilizing reported nanoprecipitation strategy, for the AIE dots (**10**) using probe (**8**), lipid PEG such as DSPE-PEG2000 and DSPE-PEG2000-Mal, were utilized to encapsulate the AIE fluorogens to avoid fluorescent dots with surface maleimide groups for novel functionalization. Further the dots were adjusted with HIV-1 Tat peptide through formation between maleimide groups on the dot surface and thiol groups at the C-ends of the Tat peptide to get Tat-functionalized greendots (**10**), similarly, RT-AIE dots (**11**) also synthesized (Fig. 8A).[45]

Qin et al. synthesized DCDPP-2TPA (**12**) AIEgens based on electron-accepting group DCDPP and electron donor group TPA to form D-A architecture (Fig. 8B). The NPs (**13**) were obtained by encapsulating probe

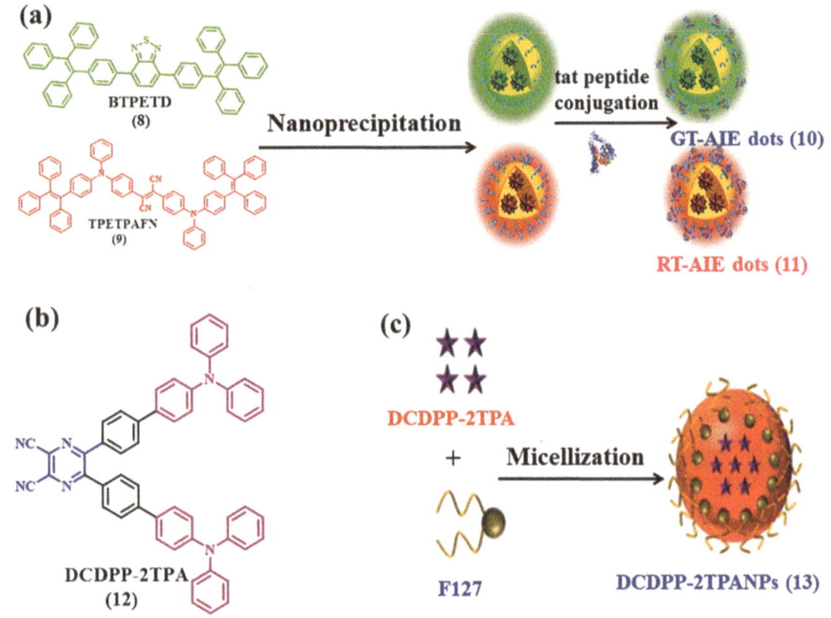

Fig. 8 (A) Schematic presentation of development of AIE dots using a probe (**8**) and (**9**); (B) Molecular structure of Probe (12); (C) stepwise illustration of the development of AIE active NPs (**13**) using a probe (**12**). *Panel (A) was reproduced and modified with permission from Li K, Zhu Z, Cai P, Liu R, Tomczak N, Ding D, Liu J, Qin W, Zhao Z, Hu Y, Chen X, Tang BZ, Liu B. Organic dots with aggregation-induced emission (AIE dots) characteristics for dual-color cell tracing.* Chem Mater. *2013;25:4181–4187. Copyright @ 2013, American Chemical Society, Panel (B–C) were reproduced and modified with permission from Wang Y, Chen M, Alifu N, Li S, Qin W, Qin A, Tang BZ, Qian J. Aggregation-induced emission luminogen with deep-red emission for through-skull three-photon fluorescence imaging of mouse.* ACS Nano. *2017;11:10452–10461. Copyright @2017, American Chemical Society.*

(**12**) with amphiphilic polymer (F127) through the micellization method (Fig. 8C). PL spectra of probe (**12**) were measured in THF and water mixture with different percentage of water fraction in vol%. As the water fraction was increased up to 40%, the fluorescence of probe (**12**) decreased due to the TICT effect. Whereas fraction of water increased between 50% and 90%, aggregation formed due to the RIM effect.[46]

The 1-PM spectrum of the NP (**13**) was obtained, in which absorption of NP (**13**) occurred in the blue and violet region. Whereas 2-PM spectrum of the NP (**13**) occurred in bright deep red emission, which suffered less photon scattering and provided deep penetration depth. The fluorescence response at excitation of 1550 fs, the 3-PM spectrum of probe (**13**) was obtained into NIR range with deep penetration depth.

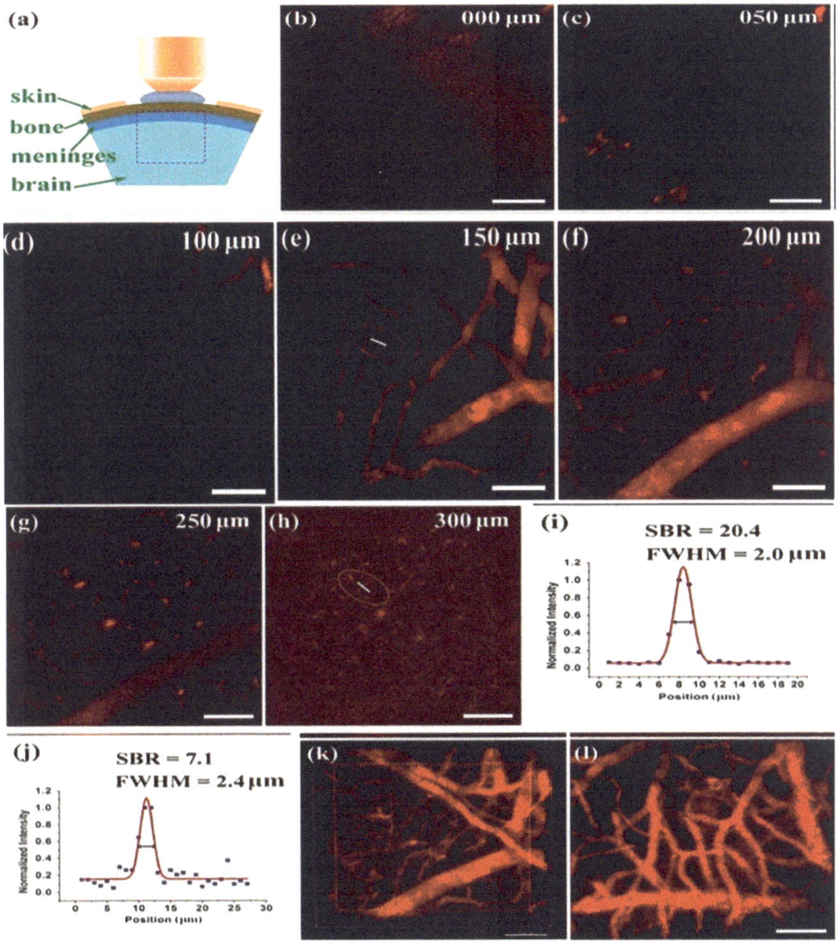

Fig. 9 In vivo 3-PM images of the brain blood vessels on a mice (A) Schematic representation of different mice brain images; (B – H) 3-PM blood vessel images at different vertical depths, 0–300 μm; (I and J) Values of fwhm of blood vessels and SBR analysis of 3-PM images at 150 μm (I) and 300 μm (J); (k) 3D architecture of the blood vessels in the mice brain, 150–300 μm; (L) 3-PM image of the blood vessels at another area in mice brain. *Reproduced with permission from Wang Y, Chen M, Alifu N, Li S, Qin W, Qin A, Tang BZ, Qian J. Aggregation-induced emission luminogen with deep-red emission for through-skull three-photon fluorescence imaging of mouse.* ACS Nano. *2017;11: 10452–10461. Copyright @ 2017, American Chemical Society.*

The in-vivo brain tissue images of mice using NP (**13**) NPs under the excitation of 1550 fs laser, observed in 3-PM spectrum. Shown in the 3-PM images can be identified up to 785 μm. The blood vessels could still be visualized with great difference at deep penetration (Fig. 9A). 3-PM images of the vertical sections of the brain blood vessels at different depths were displayed as illustrated in Fig. 9B–H. The depth of imaging was higher than 100 μm, a lot of blood vessels were noticed (Fig. 9D – F). The full width at half-maximum (FWHM) of the capillary was determined to be 2.0 μm (Fig. 9I), showing that blood vessel perceived at 150 μm underneath the skull. In the meantime, the sign-to-background ratio (SBR) was determined to be pretty much as high as 20.4, which would guarantee great imaging contrast. From Fig. 9H and J, the FWHM value was found to be 2.4 μm, with a SBR of 7.1 obtained at the depth of 150 μm. The 3D architecture of brain tissues was developed (Fig. 9K) and some selective mouse brain areas shown in (Fig. 9I).[46]

6. AIE active conjugated polymer probe for tissue imaging

Contrasted with the small molecular probe, little consideration has been paid to polymers and oligomers that show AIE, in spite of having superb properties, for example, high emission efficiency in aggregate and solid states, and the availability of different functionalization sites. Along with these highlights, if the molecular design is fully conjugated, intramolecular electronic forces between the fluorophores may show up new photophysical properties. Thus AIE-active conjugated polymers and supramolecular polymer have high fluorescence efficiencies, and they have great application in bioimaging.[47,48]

Jayakannan and *co*-worker reported oligo-phenylenevinylene based π conjugated Amphiphilic polymer for cancer tissue imaging.[49] Xia et al. reported DNA-conjugated amphiphilic AIE probe (TPE-R-DNA) cancer tissue imaging. They have developed TPE-R-DNA through hydrophobic TPE-R-N_3 group and hydrophilic single strand DNA.[50] Park and coworker reported aggregation induced enhanced emission (AIEE) material made up of biopolymeric amphiphiles having a hydrophilic and hydrophobic group such as glycon chitosan and dipolar tricyanostilbene derivatives. It is used for in vivo tissue imaging.[51]

6.1 AIE active conjugated polymeric micelle formation for tissue imaging

In 2019 Yang et al. reported pH and redox-sensitive AIE conjugated polymeric micelle for the cells and tissue imaging in the tumor. Due to the acidic and redox environment of cells, it is important to develop a pH and redox

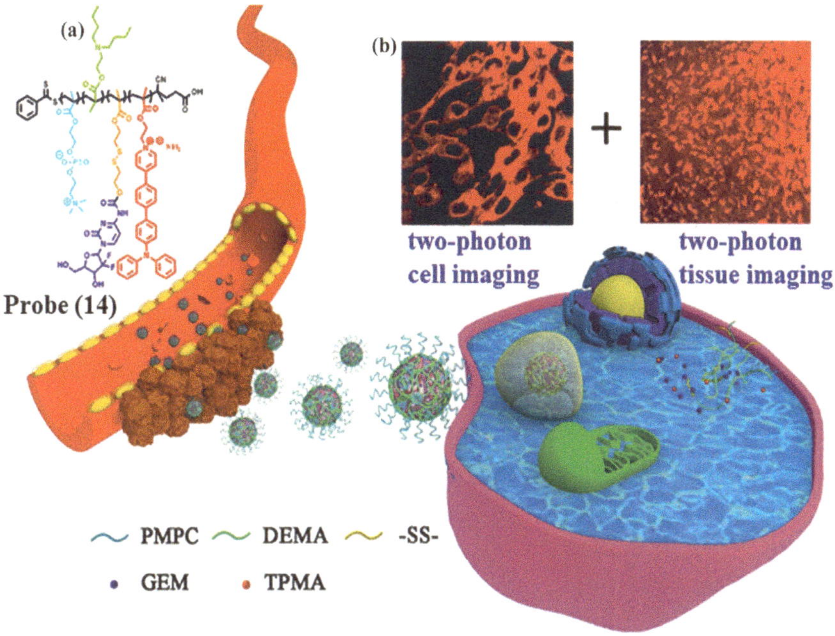

Fig. 10 (A) Stepwise illustration of conjugated micelle formation using probe (**14**); (B) Cell and tissue imaging using probe (14) containing conjugated micelles. *Reproduced with permission from Yu T, Zhuang W, Su X, Ma B, Hu J, He H, Li G, Wang Y. Dual-responsive micelles with aggregation-induced emission feature and two-photon absorption for accurate drug delivery and bioimaging.* Bioconjug Chem. *2019;30: 2075–2087. Copyright @ 2019, American Chemical Society.*

AIE CP probe. The extracellular environment of tumor tissues has pH value of 6.5–7.2, which is lower than that in ordinary tissues (pH 7.4). Gemcitabine (GEM) based conjugated redox micelles based with respect to pH-responsive PMPC-b-P (DEMA-*co*-SS-GEM-co-TPMA) (**14**) copolymer was synthesized via RAFT polymerization (Fig. 10A). The AIE property of probe (**14**) was studied by mixing different water fractions in the solution of DMSO/water. The AIE impact was seen with the expansion of water percentage, which shows probe (**14**) was suitable for developing AIE micelles. These micelles were used for hepatic and renal tissue images with 2-PM and 1-PM confocal laser scanning microscopy (CLSM) (Fig. 10B). The fluorescent image of 2-PM excitation was acquired with the imaging depth upto 150 μm. The higher resolution and deeper imaging depth of 2-PM excitation make this functional micelle to be an extraordinary probe for cells and tissue imaging.[52]

7. Conclusions and future prospect

AIEgens are the developing and advanced classes of luminescent materials which are nonemissive in solutions but strongly emit luminescence upon aggregate formation. In this chapter, we summarized the latest advancement of molecular, nanoparticle and polymeric AIE-active probes for tissue imaging. With this achievement several aspects have been noticed to important for development of specific AIE probes for tissue imaging. Some examples covered in this chapter are mainly involving short wavelength AIEgens for in vivo tissue imaging. There is need to design molecular AIEgens or NP AIEgens or CP AIEgens for long wavelength which coming under FR/NIR emissive range. This range is useful for deep penetration and low autofluorescence for tissue imaging.

Acknowledgments

SVB (IICT) is thankful to The Director CSIR-IICT for providing necessary facilities and financial support under the project BRNS No. 58/14/01/2020-BRNS/37047 IICT Commun. No. IICT/Pubs./2021/128. M.R.B. thanks to UGC, New Delhi for senior research fellowship.

References

1. Wolfbeis OS. An overview of nanoparticles commonly used in fluorescent bioimaging. *Chem Soc Rev*. 2015;44:4743–4768.
2. Förster T, Kasper K. Ein konzentrationsumschlag der fluoreszenz. *Z Phys Chem*. 1954;1:275–277.
3. Mei J, Leung NLC, Kwok RTK, Lam JWY, Tang BZ. Aggregation-induced emission: together we shine, united we soar! *Chem Rev*. 2015;115:11718–11940.
4. Hong Y, Lam JWY, Tang BZ. Aggregation-induced emission. *Chem Soc Rev*. 2011; 40:5361–5388.
5. Luo J, Xie Z, Xie Z, et al. Aggregation-induced emission of 1-Methyl-1, 2,3,4,5-pentaphenylsilole. *Chem Commun*. 2001;18:1740–1741.
6. Chen X, Wang F, Hyun JY, et al. Recent progress in the development of fluorescent, luminescent and colorimetric probes for detection of reactive oxygen and nitrogen species. *Chem Soc Rev*. 2016;45:2976–3016.
7. Garnier D, Milsom C, Magnus N, et al. Role of the tissue factor pathway in the biology of tumor initiating cells. *Thromb Res*. 2010;125:S44–S50.
8. Nam SY, Ricles LM, Suggs LJ, Emelianov SY. Imaging strategies for tissue engineering applications. *Tissue Eng Part B Rev*. 2015;21:88–102.
9. Yang B, Zhang X, Zhang X, Huang Z, Wei Y, Tao L. Fabrication of aggregation-induced emission based fluorescent nanoparticles and their biological imaging application: recent progress and perspectives. *Mater Today*. 2016;19:284–291.
10. Wu M, Junker D, Branca RT, Karampinos DC. Magnetic resonance imaging techniques for brown adipose tissue detection. *Front Endocrinol (Lausanne)*. 2020;11:421.
11. Kreitz S, Dohmen G, Hasken S, Schmitz-Rode T, Mela P, Jockenhoevel S. Nondestructive method to evaluate the collagen content of fibrin-based tissue engineered structures via ultrasound. *Tissue Eng Part C Methods*. 2011;17:1021–1026.

12. Miller S. Portable device for live tissue imaging. *Anal Chem.* 2006;78:17.
13. Spraggins JM, Schwamborn K, Heeren RMA, Eberlin LS. The importance of clinical tissue imaging. *Clin Mass Spectrom.* 2019;12:47–49.
14. Ma H, Yang M, Zhang C, et al. Aggregation-induced emission (AIE)-active fluorescent probes with multiple binding sites toward ATP sensing and live cell imaging. *J Mater Chem B.* 2017;5:8525–8531.
15. Jiang Y, Pu K. Multimodal biophotonics of semiconducting polymer nanoparticles. *Acc Chem Res.* 2018;51:1840–1849.
16. Hong G, Diao S, Antaris AL, Dai H. Carbon nanomaterials for biological imaging and nanomedicinal therapy. *Chem Rev.* 2015;115:10816–10906.
17. Hu R, Leung NLC, Tang BZ. AIE macromolecules: Syntheses, structures and functionalities. *Chem Soc Rev.* 2014;43:4494–4562.
18. Han T, Xi W, Wang D, Tang BZ. Functional polyer and systems with aggregation induced emission and stimuli response. *Top Curr Chem (Z).* 2021;379:7.
19. Niu G, Zhang R, Shi X, et al. AIE luminogens as fluorescent bioprobes. *Trends Anal Chem.* 2020;123:115769.
20. Huang X, Zhang R, Chen C, Kwok RTK, Tang BZ. Wash-free detection and bioimaging by AIEgens. *Mater Chem Front.* 2021;5:723–743.
21. Li J, Pu K. Development of organic semiconducting materials for deep-tissue optical imaging, phototherapy and photoactivation. *Chem Soc Rev.* 2019;48:38–71.
22. Miller DR, Jarrett JW, Hassan AM, Dunn AK. Deep tissue imaging with multiphoton fluorescence microscopy. *Curr Opin Biomed Eng.* 2017;4:32–39.
23. Becker JS, Breuer U, Hsieh HF, et al. Bioimaging of metals and biomolecules in mouse heart by laser ablation inductively coupled plasma mass spectrometry and secondary ion mass spectrometry. *Anal Chem.* 2010;82:9528–9533.
24. Anderton CR, Gamble LJ. Secondary ion mass spectrometry imaging of tissues, cells, and microbial systems. *Micros Today.* 2016;24:24–31.
25. Reyzer ML, Caldwell RL, Dugger TC, et al. Early changes in protein expression detected by mass spectrometry predict tumor response to molecular therapeutics. *Cancer Res.* 2004;64:9093–9100.
26. Van Nuffel S, Quatredeniers M, Pirkl A, et al. Multimodal imaging mass spectrometry to identify markers of pulmonary arterial hypertension in human lung tissue using MALDI-ToF, ToF-SIMS, and hybrid SIMS. *Anal Chem.* 2020;92:12079–12087.
27. Eberlin LS, Ferreira CR, Dill AL, Ifa DR, Cheng L, Cooks RG. Nondestructive, histologically compatible tissue imaging by desorption electrospray ionization mass spectrometry. *Chembiochem.* 2011;12:2129–2132.
28. Helmchen F, Denk W. Deep tissue two-photon microscopy. *Nat Methods.* 2005;2:932–940.
29. Feuchtinger A, Walch A, Dobosz M. Deep tissue imaging: a review from a preclinical cancer research perspective. *Histochem Cell Biol.* 2016;146:781–806.
30. Blutke A, Sun N, Xu Z, et al. Light sheet fluorescence microscopy guided MALDI-imaging mass spectrometry of cleared tissue samples. *Sci Rep.* 2020;10:14461.
31. Denk I, Strickler J, Webb W. Two-photon laser scanning fluorescence microscopy. *Science.* 1990;248:73–76.
32. Li C, Chen G, Zhang Y, Wu F, Wang Q. Advanced fluorescence imaging technology in the near-infrared-ii window for biomedical applications. *J Am Chem Soc.* 2020;142:14789–14804.
33. Dang X, Gu L, Qi J, et al. Layer-by-layer assembled fluorescent probes in the second near-infrared window for systemic delivery and detection of ovarian cancer. *Proc Natl Acad Sci U S A.* 2016;113:5179–5184.
34. Tedford CE, Delapp S, Jacques S, Anders JJ. Quantitative analysis of transcranial and intraparenchymal light penetration in human cadaver brain tissue. *Lasers Surg Med.* 2015;47:312–322.

35. Wang R, Li X, Zhou L, Zhang F. Epitaxial seeded growth of rare-earth nanocrystals with efficient 800 nm near-infrared to 1525 nm short-wavelength infrared down-conversion photoluminescence for in vivo bioimaging. *Angew Chem Int Ed.* 2014;53: 12086–12090.
36. Liu HW, Xu S, Wang P, et al. An efficient two-photon fluorescent probe for monitoring mitochondrial singlet oxygen in tissues during photodynamic therapy. *Chem Commun.* 2016;52:12330–12333.
37. Yang X, Wang Q, Hu P, et al. Achieving remarkable and reversible mechanochromism from a bright ionic aiegen with high specificity for mitochondrial imaging and secondary aggregation emission enhancement for long-term tracking of tumors. *Mater Chem Front.* 2020;4:941–949.
38. Tian H, Li D, Tang X, et al. Efficient red luminogen with aggregation-induced emission for: in vivo three-photon brain vascular imaging. *Mater Chem Front.* 2020;4:1634–1642.
39. Gao H, Bao P, Dai S, et al. Far-red/near-infrared emissive (1,3-dimethyl)barbituric acid-based aiegens for high-contrast detection of metastatic tumors in the lung. *Chem Asian J.* 2019;14:871–876.
40. Wang J, Li J, Wang L, Han T, Wang D, Tang BZ. AIEgen-based polymer nanocomposites for imaging-guided photothermal therapy. *ACS Appl Polym Mater.* 2020;2:4306–4318.
41. Gao H, Zhao X, Chen S. AIEgens-based fluorescent nanomaterials: fabrications and biological applications. *Molecules (Basel, Switzerland).* 2018;23:419.
42. Alifu N, Zebibula A, Qi J, et al. Single-molecular near-infrared-ii theranostic systems: ultrastable aggregation-induced emission nanoparticles for long-term tracing and efficient photothermal therapy. *ACS Nano.* 2018;12:11282–11293.
43. Qin W, Ding D, Liu J, et al. Biocompatible nanoparticles with aggregation-induced emission characteristics as far-red/near-infrared fluorescent bioprobes for in vitro and in vivo imaging applications. *Adv Funct Mater.* 2012;22:771–779.
44. Liu C, Wang X, Liu J, et al. Near-infrared aie dots with chemiluminescence for deep-tissue imaging. *Adv Mater.* 2020;32:1–6.
45. Li K, Zhu Z, Cai P, et al. Organic dots with aggregation-induced emission (AIE dots) characteristics for dual-color cell tracing. *Chem Mater.* 2013;25:4181–4187.
46. Wang Y, Chen M, Alifu N, et al. Aggregation-induced emission luminogen with deep-red emission for through-skull three-photon fluorescence imaging of mouse. *ACS Nano.* 2017;11:10452–10461.
47. Sánchez-Ruiz A, Sousa-Hervés A, Tolosa Barrilero J, Navarro A, Garcia-Martinez JC. Aggregation-induced emission properties in fully π-conjugated polymers, dendrimers, and oligomers. *Polymers.* 2021;13:213.
48. Li B, He T, Shen X, Tang D, Yin S. Fluorescent supramolecular polymers with aggregation induced emission properties. *Polym Chem.* 2019;10:796–818.
49. Narasimha K, Jayakannan M. Color-tunable amphiphilic segmented π-conjugated polymer nano-assemblies and their bioimaging in cancer cells. *Macromolecules.* 2016;49: 4102–4114.
50. Wang X, Dai J, Min X, et al. DNA-conjugated amphiphilic aggregation-induced emission probe for cancer tissue imaging and prognosis analysis. *Anal Chem.* 2018;90: 8162–8169.
51. Lim CK, Ki S, Kwon IC, Ahn CH, Park SY. Dye-condensed biopolymeric hybrids: chromophoric aggregation and self-assembly toward fluorescent bionanoparticles for near infrared bioimaging. *Chem Mater.* 2009;21:5819–5825.
52. Yu T, Zhuang W, Su X, et al. Dual-responsive micelles with aggregation-induced emission feature and two-photon absorption for accurate drug delivery and bioimaging. *Bioconjug Chem.* 2019;30:2075–2087.

CHAPTER TWO

AIE materials for cancer cell detection, bioimaging and theranostics

Sudesna Chakravarty[a], Sayan Roy Chowdhury[b], and Sudip Mukherjee[c,*]

[a]Department of Biochemistry and Molecular Biology, University of Nebraska Medical Center (UNMC), Omaha, Nebraska, United States
[b]Department of Chemistry, Virginia Commonwealth University, Richmond, VA, United States
[c]Department of Bioengineering, Rice University, Houston, TX, United States
*Corresponding author: e-mail address: sudip.mukherjee@rice.edu

Contents

Abstract

AIE materials exhibit weakly emissive or non-emissive properties in dilute solutions while emit powerful fluorescence in the aggregated/solid state. Recently, AIE based materials have gained immense attention due to their multifunctional role in cancer cell detection, bioimaging and cancer theranostics. In this present book chapter, we will highlight recent advancements of AIE materials for different cancer theranostics applications.

Progress in Molecular Biology and Translational Science, Volume 185
ISSN 1877-1173
https://doi.org/10.1016/bs.pmbts.2021.07.032

Graphical abstract

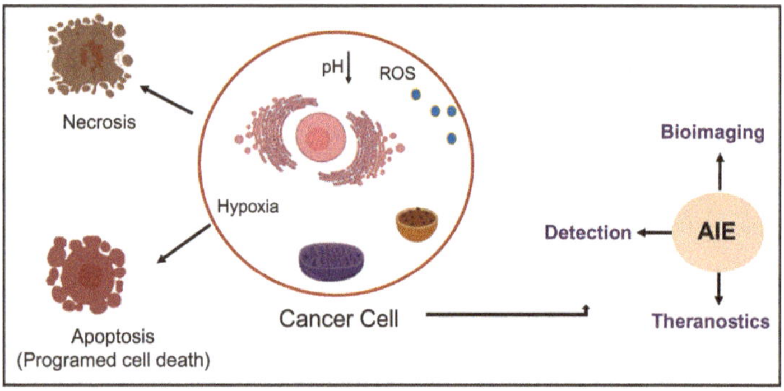

1. Introduction

Cancer is a global cause of mortality and morbidity. Cancer remains one of the leading threats to human and cause almost 9.6 million deaths in 2018. About one sixth of the total deaths worldwide occurs due to cancer. Mainly, four main lifestyle and food habit associated concerns are accountable for majority of the deaths from cancer including (i) tobacco and alcohol consumption use, (ii) lack of physical action, (iii) high-level body mass index and iv) low vegetable and fruit consumption. Additionally, viral infections (~25%) is another leading reason for causing cancers to humans mainly in poverty-stricken countries.[1,2] Conventional cancer therapies including chemotherapy, radiation, surgical intervention, chemotherapy, either alone or in a combination, has continued to be the preferable treatment modalities over years.[3] However, all these conventional treatment modalities have several challenges including drug resistance, toxicity, unwanted side effects etc. Moreover, tissue biopsy, MRI or X-ray had been commonly used techniques for the prognosis of cancer cells. However, in recent year several new techniques have been emerged as an alternative and effective route including positron emission tomography (PET), ultrasound (US), X-ray (plain film and computed tomography [CT]), optical imaging, and single-photon emission computed tomography (SPECT).

In recent years, an improved insight of cancer biology has prompted a brand new age of targeted cancer therapeutic strategies by using several unique attributes of cancerous cells.[4] Immunotherapy and nanomedicine have gained immense attention for modern cancer theranostics applications due to several advantages over their conventional counterparts. In spite of these considerable progresses, every approach suffers from around inherent restrictions which have shifted the focus on the expansion of various alternative nanomaterials and multifunctional materials including metal nanoparticles, polymeric nanoparticles, bio-nanoparticles, liposomes, lipid-polymer hybrids and AIE based materials, etc.[5–10]

In this context AIE based materials and nanoparticles has gained immense attention due to its multifunctional properties including imaging, anticancer and biosensing. AIE based materials were used for cancer theranostics in various in vitro cancer cells including MCF-7 (a human breast cancer cells that was isolated from a tumor bearing woman back in 1970), Cellosaurus cell line HLF (CVCL_2947) etc.[11,12] AIE materials exhibit inadequately emissive or non-emissive properties in the dilute conditions while emit powerful fluorescence in aggregated or solid state. The primary mechanism to interpret such AIE phenomena is the constraint of intramolecular motions (RIM), and intramolecular rotation (RIR) along with the restriction of intramolecular vibration (RIV). Further AIE materials show excited state intramolecular proton transfer (ESIPT), which in turn increases the possibility of development light sensitive probes with multiple features such as sharp imaging contrast, significant stokes shift and enhanced high signal-to-noise ratio (Fig. 1). The photophysical process involved in such AIE based luminescent materials can be explained in terms of Jablonski diagram. Such AIE materials have found wide applicability for cancer cell detection, bioimaging and theranostics applications, which in turn plays pivotal role in the cancer management. AIE materials possess distinct advantages in terms of large stokes shift, high brightness, great biocompatibility and excellent photostability.[13] Two important types of such AIE materials are tetraphenylethylene (TPE) and 1-methyl-1,2,3,4,5-pentaphenylsilole (Silole).

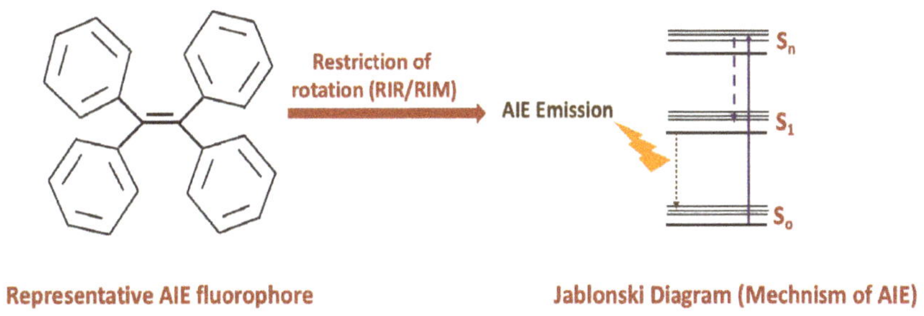

Fig. 1 Mechanistic insight of AIE phenomenon.

Here in this present book chapter, we will highlight various biomedical applications of AIE based materials in cancer cell detection, bioimaging, anti-cancer applications and theranostics. Further we will discuss the existing challenges and potential solutions for the clinical translations of these materials.

2. AIE based materials in cancer cell detection

AIE materials have been reported to play important role in designing selective and sensitive detection assays for biomolecules.[14,15] Recently, monitoring altered conditions within tumor microenvironments such acidic condition, ROS and hypoxia with AIE detection probes has emerged as an important section for both cancer diagnosis (Fig. 2). An important hypoxia AIE probe was reported by You *et al.*[16] Herein, they have detected nitro reductase quantitatively and evaluated hypoxia status of cancer cells using TPE based AIE Probe. A limit of detection (LOD) of 5 ng mL^{-1} was achieved for NTR.[16] Another AIE probe for highly selective, specific, simple and rapid in situ monitoring of CO_2 generation rate was also reported.[17] Chen *et al.* developed an AIE based TPP-TMAE probe [tris (2-(dimethylamino) ethyl)-4,4′,4″-(1H-pyrrole-1,2,5-triyl) tribenzoate)] with targeted applications toward three types of cells followed by CO_2 generation rate monitoring.[17] To be precise, the percentages of CO_2 generation in MCF-7, MEF and HeLa cells were obtained to be $5.78 \times 10^{-6} \pm 6.0 \times 10^{-8}$ μg/h, 6.40×10^{-6}-$\pm 6.0 \times 10^{-8}$ μg/h and $4.27 \times 10^{-7} \pm 4.0 \times 10^{-9}$ μg/h in that order.[17] This study shed light on the applicability of AIE probe for monitoring metabolic conditions within cancer cells, which play important role in the early-stage cancer diagnosis.[17] Behnke *et al.* reported a superior AIE probes based on asymmetric cyanine Itrybe with unique characteristic features such as absorption-emissive bands in the analytical window.[18] Further, they have shown application of the AIE probe for pre-clinical imaging applications toward tumor specific biomarker-HER 2.[18] For these nanoparticles loaded

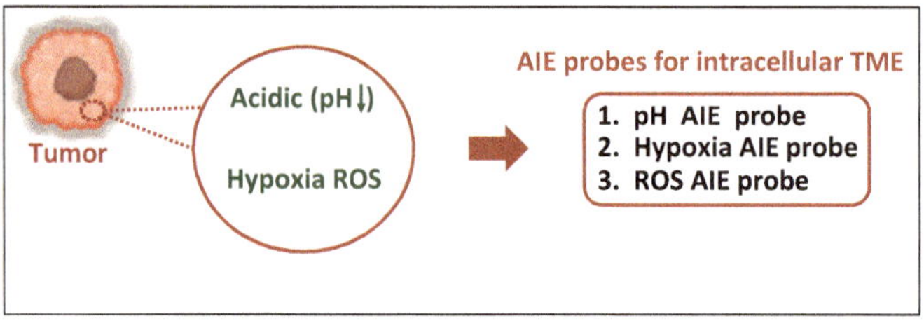

Fig. 2 Cancer detection by AIE probes.

with Itrybe were surface functionalized with PEG (polyethylene glycol) and Herceptin, an antibody that targets the tumor.[18] The applicability was shown for both in vivo and in vitro studies-immunoassays on tumor cells and sections from mouse tumor xenograft as well. It is expected for such AIE probes to act as a bright and versatile NIR probe for sensitive detection as well as characterization of tumor development/progression.[18] It is reported that ROS is implicated in the control of telomerase activity pathway in cancer cells through Akt signaling. The detection of telomerase activity plays pivotal role in the early cancer detection and diagnosis.[11] In this regard, Zhuang *et al.* reported the utilization AIE probes composed of Silole-R, TP*E*-H labeled primer (QP) and extension repeated units leading to dual fluorescence signal generation (478 and 605 nm).[11] The activity of the probe was used to uncover telomerase in vivo as well as in vitro (via mapping) from a variety of cell lines such as HeLa, HLF, E-J, MCF-7 along with clinical samples-normal (urine samples from 10 patients), bladder cancer (urine samples from 50 patients). A limit of detection (LOD) down to cells was achieved. A fundamental engineering aspect of AIE probe design was reported by Bai *et al.*[19] Specifically, they have reported the generation of supramolecular AIE probes for fluorescence-based differentiation among normal cell line, two cancer cell lines and two metastasized cell lines. The AIE probes were fabricated using supramolecular approach of host (TPE)-guest [Cucurbit[8]uril] interaction. Convincing results via utilization of such probes were reported. They showed differentiation ability within 20 min devoid of cell disruption, washing steps or centrifugation. Thus, such strategy is expected to play pivotal role in fundamental design strategy for AIE probes as well as their clinical applications.[19] A novel pH responsive AIE probe was reported by Dutta *et al.*[12] The AIE nanoprobe was based on luminescent copper nanocrystals (CuNCs) showcasing pH dependent luminescence change within the cellular environment of living cancer cells. In order to observe the intracellular kinetics, two types of cell lines were taken into account-MCF-7 and HEK-293. It was found that AIE kinetics in MCF-7 cells was 3 times more than HEK 293 cell lines. Thus, this study showed the potential of AIE in different cell lines depicting future application potential in multiple applications such as detection, bioimaging as well as theranostics.

3. AIE based materials in bioimaging

Bioimaging techniques such as super-resolution nanoscopy, multiphoton fluorescence microscopy, confocal microscopy and light-sheet ultramicroscopy have shown great potential in understanding biological events such as cell tracking, image-guided surgery, and vascular disease diagnosis.[20,21] Such imaging is an irreplaceable technique in biomedical sciences owing to multiple advantages such as high sensitivity, excellent spatiotemporal resolution, non-invasiveness, rapidity, real-time applicability, and facile accessibility.[14,15,22,23]

The imaging quality of these bioimaging techniques is however highly dependent on the nature of contrast agents. Typically, such contrast agents are based on various fluorescent materials-organic dyes, fluorescent conjugated polymers, metal nanoclusters, quantum dots, up-conversion nanoparticles, carbon nanomaterials etc. Among these, organic materials possess immense potential for real time clinical applications owing to good biocompatibility, facile processability, tunable spectral properties and versatile modification strategies. Two important subtypes of organic materials—"organic fluorophores" are termed as AIE (aggregation induced emission) and ACQ (aggregation-caused quenching).[14,15] Between these, AIE materials has attracted much attention as contrast agents, primarily as ACQ suffers from drawbacks in terms of limitation in the form of the label to analyte ratio which forces the utilization of highly dilute solutions of fluorophores compelling the utilization with a narrow scope of practical applications.[22] AIE materials have special significance in bioimaging applications as multiple washing steps can be eliminated (Fig. 3). The aggregation of AIE fluorophores were induced using ionic amphiphiles or AIE fluorophores conjugated with suitable ionic factions via intermolecular interactions including hydrophobic and electrostatic.[14,15]

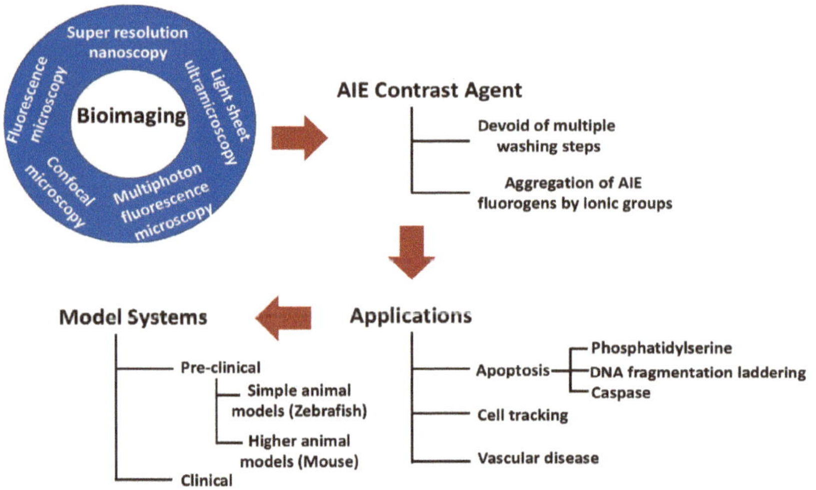

Fig. 3 Bioimaging applications of AIE materials.

In case of in vivo imaging for larger tissue penetration, AIE probes with bathrochomic shift from far red, NIR I [(700–900) nm, lowered tissue absorption] to NIR II [(1000–1700) nm, decreased photon scattering along with tissue auto-fluorescence] is preferred. To understand the applicability of such AIE luminogens, these must be used in different animal models (Simple animal models—Zebrafish) and (higher animal models—mouse). So far, different AIE bioimaging probes for cancer has been reported. An important biological phenomenon associated with different cell growth and proliferation pathways is apoptosis. The deregulation cell apoptosis has major implications in cancer particularly cancer progression, efficiency of cancer therapy, apoptosis drug screening etc. Thus, monitoring apoptosis-early and late plays key roles in cancer. There are basically three ways of apoptosis bioimaging-the monitoring of phosphatidylserine (PS), the laddering of DNA fragmentation and the monitoring of caspase activation on the apoptotic cells. An important AIE based apoptosis probe for monitoring PS was developed by Hu *et al.*[23] Specifically, they have developed an AIE apoptosis bioimaging probe based on zinc coordination and a salicylaldazine fluorophore skeleton (AIE-ZnDPA). It exhibits AIE effect for both early and late apoptosis. It particularly undergoes fluorescence "Turn On" for the apoptotic cells (early-stage) by attaching to the negatively charged membrane containing PS. In addition, when the cells go through late-stage apoptosis, the AIE bioimaging probe penetrates through cell membrane entering to nuclei followed by interaction with nucleic acid leading to fluorescence "Turn On." This probe was further utilized for hydrogen peroxide and staurosporine mediated cell apoptosis. Thus, it allows sub-cellular localization for biological monitoring of apoptotic process that play pivotal role in cancer. Another important target for understanding cancer cell apoptosis is caspase. Caspase is an intracellular protease enzyme which play key roles in the initiation and execution of apoptosis. Within caspases, caspase-3 is an important facilitator of cell apoptosis, which can be bioimaged by propeller shaped tetraphenylethene (TPE) based AIE fluorogens. Keeping these in mind, Shi *et al.* have designed an AIE bioimaging probe based on hydrophilic DEVD peptide (Asp-Glu-Val-Asp) sequence and a hydrophobic AIE based fluorogen.[24] Precisely, the cleavage of DEVD peptide with caspase-3/-7 causes AIE that leads to fluorescence "Turn On" in living cells. The sequence of AIE along with Photoinduced Electron Transfer (PET) is another novel design strategy for designing highly selective and sensitive AIE probe for elevated signal output and reduce background contrast to PET or AIE mechanism alone. Thus, Zhan *et al.* developed two derivatives of

TPE-Zincke's salt substituted TPE with counterions Cl^- and PF_6^- respectively, for sensing of glutathione (which plays vital role in homeotastis) and bioimaging of cancer cells.[25] Though much advancement in AIE and hybrid AIE based organic fluorophores has been observed but recently much focus has been given toward nano-based AIE probes particularly owing to multiple advances in terms of higher brightness, better photostability, larger absorption coefficients. However, in terms of clinical translation, organic fluorescent nanoparticles encapsulated inside biopolymer are considered more favorable over inorganic nanoparticles. Thus, Zhang *et al.* reported the synthesis of biopolymer coated organic fluorescent nanoparticle, PEG-lipid coated 2TPE-4E as AIE nanoprobe for cancer cell (MCF-7/A549/293T cells) targeting and bioimaging applications.[26] Another organic fluorescent nanoparticle as an AIE probe for non-invasive bioimaging applications was reported Che *et al.*[27] Herein, at first AIE luminogens were generated by attachment of TPE units to BIODIPY core (AIEgens). Thereafter, the AIE NPs were fabricated in presence of DSPE-PEG_{2000} which showed excellent properties such as good water dispersibility, biocompatibility and excellent photostability. The generated NPs were found to be able to carry out cell bioimaging applications within few seconds via endocytosis of cells. This is particularly important in cancers bearing solid tumors such as PDAC, breast cancer etc. A. Nicol and co-workers demonstrated the fabrication of a new AIEgen having huge potential for solid state bioimaging resulting from a favorable absorption as well as deep-red emission.[28] Specifically, they have prepared TPE-TETRAD as AIEgen which possess two fundamental outstanding properties-aggregation induced three photon luminescence (non-linear optical property) and a very high two photon absorption cross-section (313 MG at 830 nm). The fabricated AIEgen showed low cytotoxicity and low ROS generation potential, higher photostability and demonstrated a high potential to perform multiple applications such as mitochondria imaging, multiphoton imaging etc. Similarly, another AIE probe was fabricated for bioimaging in vitro and in vivo models.[29] It was based on donor-acceptor complex (TPA-DCPE; 1,1-dicyano-2-phenyl-2-(4-diphenylamino)phenylethylene]) followed by encapsulation with pluronic F-172 to develop NPs that show three photon fluorescence signal (3PF). A quantum yield of 29.3% with excellent biocompatibility was achieved for TPA-DCPE based nanoparticles. This NP was used in the cellular imaging of HeLa cells demonstrating bright red fluorescence in cytoplasm. Further, it was used for in vivo 3PF imaging under 1560 nm fs laser resulting in a deep tissue penetration deepness of 300 μm with a high-level spatial resolution of

1.8 μm with SBR 14 in a 3D model of vasculature underneath the skull. This work provides an important outlook for the construction of efficient 3PA AIE probes. Continuous in vivo/in vitro cell tracing in a non-invasive manner plays pivotal role in a variety of biological processes such as fundamental understanding of genesis, invasion, development, cancer metastasis, tissue regeneration monitoring post stem cell transplantation etc.[30] Herein, Li *et al.* fabricated AIE active TPETPAFN and FR/NIR AIE dots functionalized by peptides (cell penetrating) for in vivo/in vitro cell tracing applications. Typically, such AIE dots exhibit outstanding properties such as high photostability and size, brighter fluorescence, better cell tracing ability, brighter fluorescence ability, strong photobleaching resistance and excellent biocompatibility compared to inorganic quantum dots.[22,30] Further an AIE based dual modal MRI contrast agent was prepared by Chen *et al.* The fabricated AIE probe comprised of TPE fluorophore and two Gd diethylenetriaminepentaacetic acid (hydrophilic) moieties (TPE-2Gd).[31] It was used as a fluorescence cell imaging agent with excellent photostability and low to nil cytotoxicity. In terms of MRI capability, TPE-2Gd exhibits a longitudinal relaxivity in water comparable to commercial agents. It is found to have an increase in circulation lifetime from 10 min to 1 h and great renal filtration. Thus, it exhibits the clinical ability for the use of a MRI contrast agent for liver imaging.

4. AIE based materials in theranostics

Recently AIE probes has also been exploited for cancer theranostics applications (Fig. 4). Such applications play important role in cancer therapy-cell organelle targeted drug delivery, image guided laser ablation, image guided surgery etc. The specific cell organelle targeted anti-cancer drug delivery leads to enhancement in therapeutic efficiency. Thus, Feng *et al.* has reported an AIE probe for subcellular organelle-mitochondria

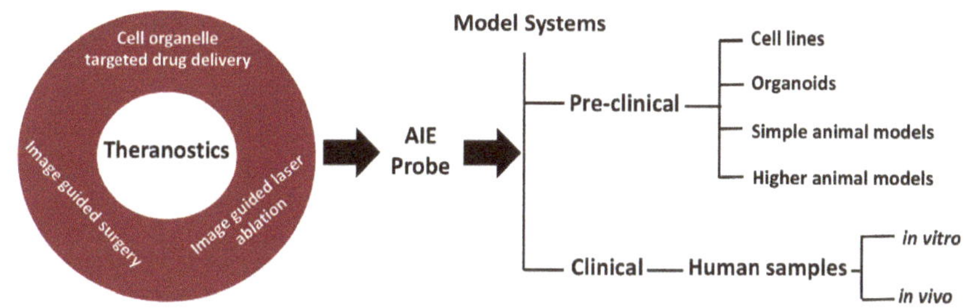

Fig. 4 Theranostic applications of AIE probes.

targeting along with imaging application (image guided cancer cell ablation).[20,21] To be precise, they have reported image guided cancer cell ablation utilizing TPE thiophene (TPETH) [TPETH-Mito-IART) for the codelivery of artemisinin (ART) and AIE photosensitizer to cell specific mitochondria. ROS is generated upon ART accumulation in cancer cell mitochondria (which in turn induces PDT performance) with the simultaneous fluorescence "Turn on." The phenomena depolarizes mitochondrial membrane and reduces cancer migration ability.[20,21] Thus, the process of generation and application of AIE theranostic probe helped in cancer cell ablation. Similarly, Ma *et al.* reported an AIE based theranostics platform for targeted drug delivery and tumor retention.[32] Specifically, a two-photon skeleton (AIE and large two-photon absorption cross-section) was constructed using conjugated polymer P (TPMA-*co*-AEMA-PEI (DA)-Blink-PEG ($PAEE_{blink}$-DA) containing therapeutic payload-Doxorubicin (Dox). It exhibited tissue penetration depth upto 150 μm along with the capability of in vivo biodistribution tracing of NPs. In another work, Han *et al.* also created a template formulation for anti-cancer drug Doxorubicin (Dox) delivery monitoring.[33] Specifically, TPE/Dox encapsulated by pH sensitive amino poly(glycerol methacrylate) derivative (PGMA-EDA) was generated for controlled intracellular release of Dox. They visualized targeted tumor delivery by the attachment of folic acid ligands at the side chain of PGMA-EDA. Herein, they have utilized two photon excitation technique which has high potential in terms of elevated contrast, penetration to deep tissue as well as lower toxicity.[33] Another theranostic platform was developed by Dai *et al.*[34] Specifically, they have developed NIR emissive water soluble fluorescent AIE probe for dual function-bioimaging of organelle as well as PDT functions. They have shown bioimaging potential by long term tracking of lysosomes in living cells as well as zebrafish with an overlap coefficient of 0.9687. In addition, PDT ablation of cancer cells was observed under green light irradiation. This study expands the applicability of NIR AIE probes for the treatment of cancer. Another AIE theranostic platform was synthesized based on donor acceptor complex for the applicability as far-red/NIR photosensitizer for image guided photodynamic therapy.[35] Specifically, the donor-PTPEAQ was synthesized via one-pot Suzuki polymerization followed by functionalization with anti-Her 2 antibodies. It showed FR/NIR emission at 680 nm along with higher quantum yield ($^1O_2$) as well as better photostability compared to commercial photosensitizer-Rose Bengal. The photodynamic cancer ablation was visualized through a breast cancer line SKBR-3 with HER2 overexpression. Jin and coworkers showed the fabrication of UCNP@TTD-cRGD NPs exhibiting AIE properties for selective NIR PDT of deep-seated tumors.[36] The stated NPs can efficiently generate ROS in 6 mm thick tissue upon 980 nm laser excitation. Further, it has shown the potential of efficient and selective targeted killing of MDA-MB-231, a triple negative breast cancer cells as well 2D/3D in vitro tumor models. AIE nanodots as cancer theranostic platform has also been investigated. Gao *et al.* reported the design and fabrication of red emissive AIE nanodots for image guided PDT by targeting integrin $\alpha_v\beta_3$.[37,38] The synthesized AIE nanodots exhibited multiple advantages such as elevated brightness, profound ROS production capability, excellent biocompatibility with devoid of significant toxicity. The targeted efficiency was checked using HCC cell xenograft tumor model. Thus, it provides a practical platform for image-guided PDT. Apart from these, AIE dots has also been explored for image guided surgery. Ni *et al.* showed the development of afterglow luminescence (AGL) AIE dots as AIE probes that has the potential of exhibiting luminescence over 10 days after stopping the single excitation, which is particularly helpful for image guided surgery.[39] The applicability of AGL AIE dots was shown via post-intravenous injection of AGL AIE nanodots into mice in the peritoneal carcinomatosis model. It showed 100-fold higher AFL imaging of tumor to liver ratio compared to fluorescence imaging. This showed a huge potential AGL AIE dots for image guided cancer surgery.

5. AIE based materials for anticancer application

After the discovery of AIE concept in 2001, several advancements have been made in order to utilize light for theranostic applications outsmarting the traditional fluorescence materials in anticancer applications. Significant progresses have been achieved to improve the AIEgens in terms of higher photostability, biocompatibility, high specificity, brightness and signal to noise ratio toward various therapeutic functions.[20,40] Light up AIE probes were empowered with theranostic capabilities via making multifunctional probes (Fig. 5).[41] AIE based nanoparticles with variable size and morphologies were developed via encapsulating AIEgens into block copolymers. These nanoparticles were utilized successfully for tumor tracking, bioimaging, sensing and image-guided surgery. Similarly, AIE dots were developed using AIE fluorogens as core and biocompatible NPs acting as a shell with sizes varying from few to tens of nanometer. They proved not only to be a promising contrast agent in animal models and as well as in several imaging techniques but also show a linear loading percentage-dependent increase in brightness. After introducing AIEgens into the photosensitizers, non-radioactive decay was controlled in the aggregated state to boost two competing processes photosensitization and light emission. Introducing donor−acceptor pairs and strength energy gap between T_1 and S_1 can be

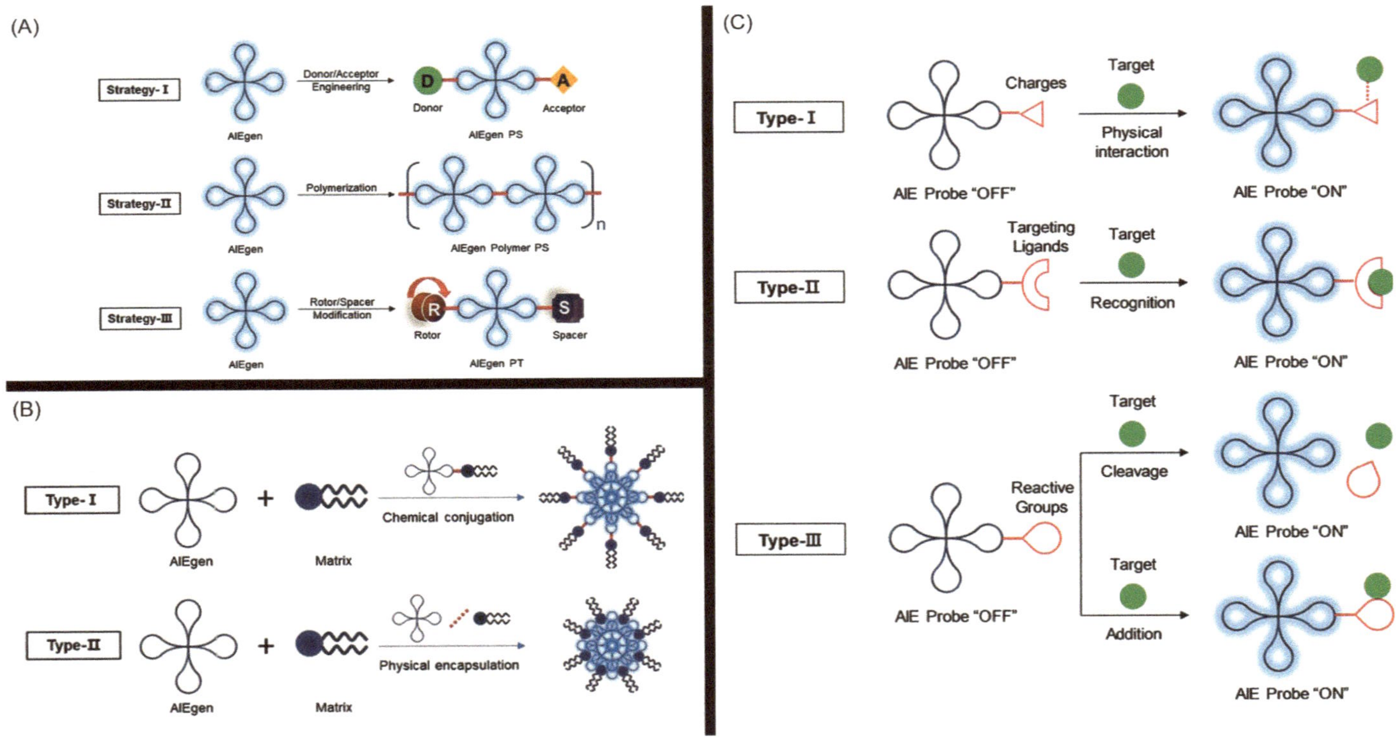

Fig. 5 Design strategies of (A) AIE based molecular probes (B) AIE Nanoparticles (NPs) and (C) AIEgens with theranostic functions. *Reproduced with permission from Cai X, Liu B. Aggregation-induced emission: recent advances in materials and biomedical applications.* Angew Chem Int Ed Engl. *2020;59:9868–9386. Copyright 2020 Wiley-VCH.*

modified, leading to efficient intersystem crossing (ISC) which is essential for efficient ROS generation.[42] A small ΔE_{st} can lead to efficient photosensitization.[43] For two photon photodynamic therapy, AIE dots with large two photon absorption cross sections and efficient ROS production serves as the most promising agents.[44] Mitochondria targeted probes are the most common find as it is the main target of ROS.[45] Chemiexcited platform was introduced to overcome the limitation of penetration depth of PDT therapy and it produced in brighter emission in animal models without any background signal realizing power free image guided therapy unlike conventional systems.[46] The key to make AIEgens of high photothermal conversion potential was to choose the right rotors and spacers into the AIE skeleton that can promote non radiative pathways. In this section we choose to focus the AIE materials (Fig. 6) that have been made in recent years for anticancer applications (Table 1).

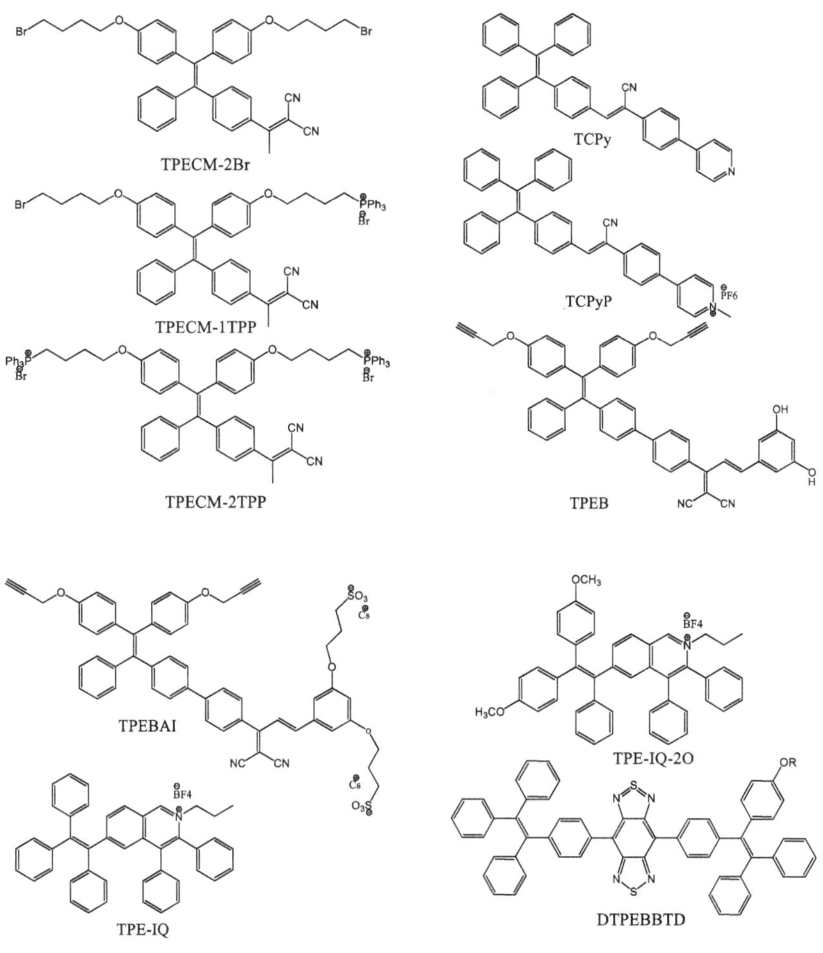

Fig. 6 Chemical structures of different precursor AIE materials.

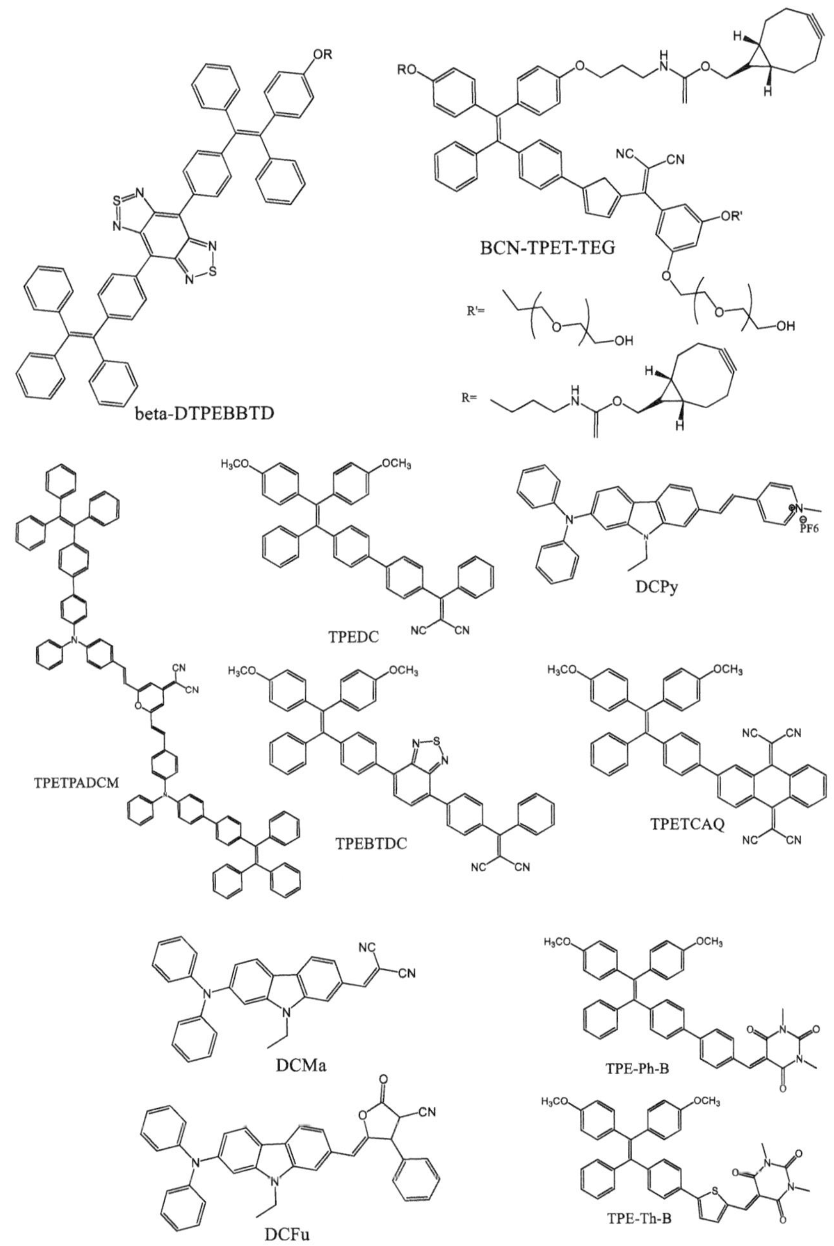
beta-DTPEBBTD
BCN-TPET-TEG
TPETPADCM
TPEDC
DCPy
TPEBTDC
TPETCAQ
DCMa
TPE-Ph-B
DCFu
TPE-Th-B

Fig.6—Cont'd

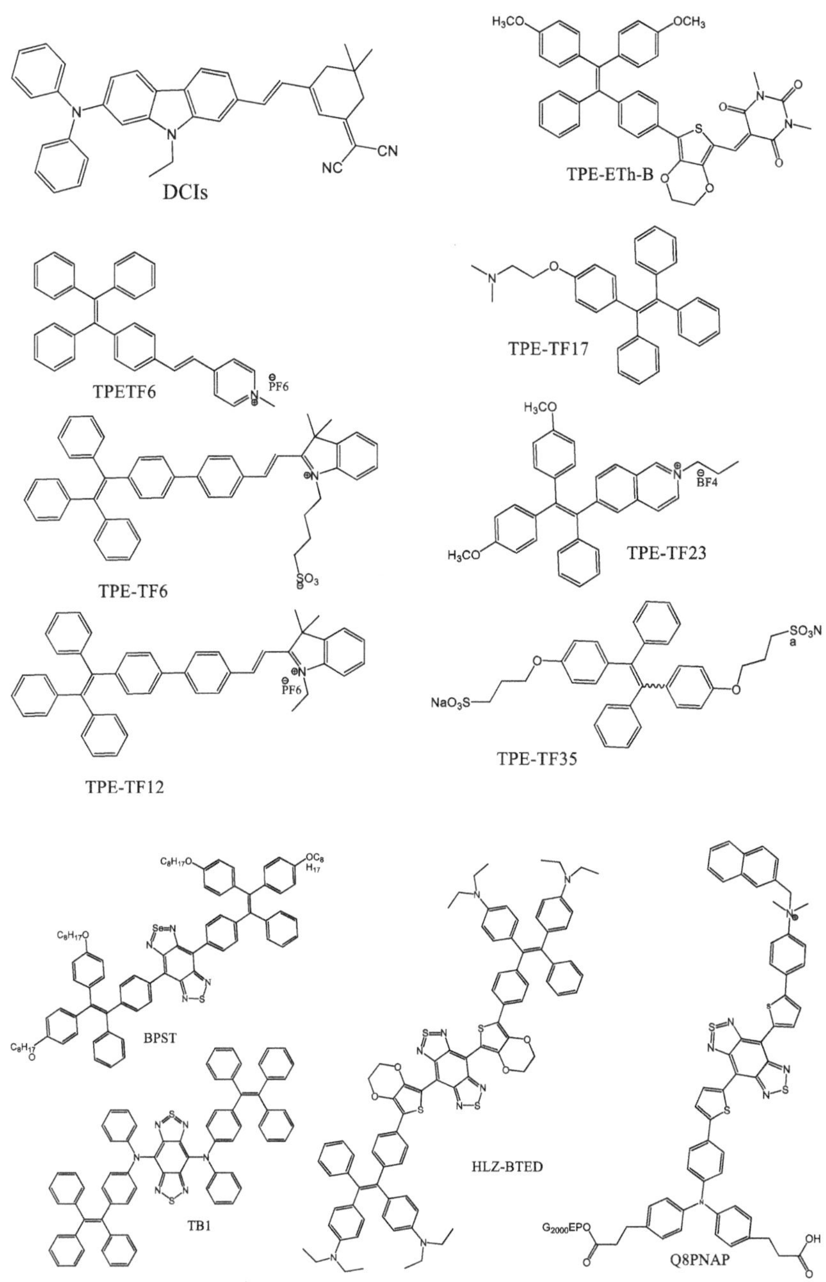
DCIs
TPE-ETh-B
TPETF6
TPE-TF17
TPE-TF6
TPE-TF23
TPE-TF12
TPE-TF35
BPST
TB1
HLZ-BTED
Q8PNAP

Fig.6—Cont'd

Table 1 AIE materials for anti-cancer application.

Sr. no	Type	Material	Skeleton	Advantages/disadvantages	Application	References
1	Mitochondria targeting	TPE-IQ	TPE and isoquinolinium-based molecule	Low specific attachment	PDT ablation of cancer cell	47
2	Mitochondria targeting	TPECM-1TPP, TPECM-2TPP, TPECM-2Br	TPE and triphenylphosphonium based molecule	Tumor microenvironment-activated drug release, prodrug for therapy	"Turn On" image guide at Near IR and therapeutic effects	14,15
3	Mitochondria targeting	TPE-IQ-2O	Modified TPE-IQ with methoxy groups	Kills cancer cells through ROS generation	Photosensitizer	48
4	Mitochondria targeting	TCPy, TCPyP, PEG-AIE-TPP micelle	TPE and phenylacrylonitrile in TCPy and TCPyP, hydrazone containing the copolymer	long-term tracing of the tumor in mice	Cell imaging, sensing, cancer diagnosis, anti-counterfeiting	49
5	Mitochondria targeting	AIE-mitoprobes	Traditional TPE based systems	Drug delivery, PDT performance when photosensitizers are included	Cancer cell selective cytotoxicity	45
6	Cell membrane targeting	BCN-TPET-TEG, cRGD-S-Ac3ManNAz and TPBAI	AIE core (TPET), hydrophilic triethylene glycol (TEG), and copper-free clickable bicyclo[6.1.0] nonyne (BCN) alkyne functionalized TPEBAI and a dual-responsive metabolic precursor, cRGD–S-Ac3ManNAz	Rapid tumor specific imaging, biorthogonal labeling	In vivo biorthogonal labeling, amplified PDT performance compared to mitopprobes	50

7	In vitro imaging	PLGA dots loaded with TPETPAFN	2,3-Bis(4-(phenyl(4-(1,2,2-triphenylvinyl)phenyl)amino)-phenyl)fumaronitrile (TPETPAFN) loaded in PLGA NPs	MCF-7 cells show bright red emission	Targeted cellular imaging	51
8	In vitro Mitochondria imaging	AIE dots with folate and triphenylphosphine	Nanoprecipitation method was followed lipid-PEG as encapsulation matrix bearing folic acid and triphenylphosphine (TPP) at the surface	Selectively internalized into MCF-7 cells	Mitochondria labeling of MCF-7 cells	52
9	Image guided cancer surgery	(DTPEBBTD) dots	Donor–acceptor–donor (D–A–D) structure, monoalkoxy-substituted TPE as a relatively weak electron donor with benzo[1,2-c:4,5-c′]bis([1,2,5]thiadiazole) (BBTD) as acceptor	Removal of tumors with real time visualization	distinctly differentiate tumor nodules compared to normal tissues	53
10	Sentinel Lymph Node (SLN) imaging	TPETTQ dots	Two propeller structures of tetraphenylethene (TPE) units as the electron donor to 4,9-di-(5-bromothiophen-2-yl) thiadiazolo-quinoxaline (TTQ) as electron acceptor, D-A-D type structure	High-resolution Photo acoustic imaging of SLN and good photothermal conversion efficiency	Photoacoustic Imaging along with Photothermal removal of Cancer Cells	54

Continued

Table 1 AIE materials for anti-cancer application.—cont'd

Sr. no	Type	Material	Skeleton	Advantages/disadvantages	Application	References
11	Targeted PDT cancer therapy	TPETPADCM-cRGD dots	Encapsulation of 2-(2,6-bis((E)-4-(phenyl(4′-(1,2,2-triphenylvinyl)-[1,1′-biphenyl]-4-yl)amino)styryl)-4H-pyran-4-ylidene)malononitrile (TTD) using 1,2-distearoyl-sn-glycero-3-phosphoethanolamine-N-[maleimide(polyethylene glycol)] (DSPE-PEG-Mal) as the encapsulation matrix, surface functionalization with cyclic arginine-glycine-aspartic acid (cRGD) tripeptide	Killed tumor cells via generating ROS, inhibits tumor growth	Accumulation at tumor site via both active and passive targeting	55
12	Photosensitizers	TPETCAQ dots, MeTTPy	Encapsulation of TPETCAQ using 1,2-distearoyl-sn-glycero-3-phosphoethanolamine-N-[maleimide(poly(ethylene glycol))-2000] (DSPE-PEG-MAL) as the matrix, surface functionalization with the HIV-1 transactivator (RKKRRQRRRC)—TPETCAQ dots strong D-A type skeleton, pyridinium moiety (A), carbon–carbon double bond (π-bridge), thiophene fragment (D and π-bridge), triphenylamine segment (D), and/or two methyl groups (D)-MeTTPy	Excellent in vivo PDT performance along with efficient photosensitization	NIR AIE emission for image-guided PDT using ultra low power lamp light	56

13	Two photon PDT treatment	TPEDC dots, TTPM based on a rigid Donor-π-Acceptor skeleton, Push-pull AIEgens: DCMa, DCIs, DCFu, DCPy	Encapsulation of TPEDC using 1,2-distearoyl-sn-glycero-3-phosphoethanolamine-N-[maleimide(poly(ethylene glycol))-2000] (DSPE-PEG-MAL) as the matrix, surface functionalization with the HIV-1 transactivator (RKKRRQRRRC)—TPETCAQ dots $D-\pi-A$ structure, triphenylamine and the pyridinium moieties were bridged through phenyl and thiophene groups to regulate the molecular electron effect-TTPM-Push Pull AIEgens electron-donating diphenylamine group is attached to four different electron-withdrawing groups (malononitrile, isophorone, 3-cyano-4-phenyl-2(5H)-furanone and methylpyridinium salt) through electron-rich carbazolyl ring-DCMa, DCIs, DCFu and DCPy	AIE dots with large Two photon absorption cross sections and high ROS production	In vivo blood-brain vessel closure and PDT for cancer cell removal	57
14	Chemiexcited PDT	C-TPEBTDC dots	Co-encapsulating TBD and bis [2,4,5-trichloro-6-(pentyloxycarbonyl)phenyl] oxalate (CPPO) into amphipathic pluronic F-127 in the presence of soybean oil	Accumulated at tumor sites to yield bright red-NIR luminescence, inhibit tumor growth via ROS generation	Utilization of localized H_2O_2 as a theranostic trigger	58

Continued

Table 1 AIE materials for anti-cancer application.—cont'd

Sr. no	Type	Material	Skeleton	Advantages/disadvantages	Application	References
15	In vivo Bio-orthogonal labeling	DBCO-AIE dots	AIE dots were prepared by nanoprecipitation method using DSPE-PEG with DBCO: DBCO-PEG-lipids was coated on the polyyne bridged red-emissive 2TPE-4E (D–π–A structure): it has two TPE units on both ends of a polyyne	Specific targeting to the tumor cells undergoing in vivo metal-free click reaction	Biorthogonal labeling for selective tumor diagnosis	42,59
16	Imaging metastatic tumors	TPE-Th-B, TPE-Ph-B, TPE-ETh-B based AIE dots	D–π–A structure: TPE as Donor and (1,3-dimethyl)barbituric acid-as acceptor dots were prepared using nanoprecipitation method and Lipid-PEG_{2000} was used as doping matrix	High resolution bioimaging of cells with high signal to noise ratio	Pulmonary metastatic tumors are detected in live animals	37,38
17	In vitro anti-cancer activity	A series of TPE-TF based AIEgens	D–π–A structure: TPE-TF6, TPE-TF11, TPE-TF12	AIEgens can differentiate between normal human cells and cancer cells	Display strong cytotoxicity, resistance to cancer invasion, effect the cell cycle, induce apoptosis along with ROS generation	60

18	Near IR-II AIEgens (Donor-Acceptor-Donor type skeleton)	BPST, TB1, HLZ-BTED, Q8PNap, 2TT-oC6B, 2TT-oC26B BTPPA, XA1, DPTQ-PTZ, P3c, HL3, HQL2, Q-OH, BPN-BBTD, BPBBT	D-A-D type structure, Benzo [1,2-c:4,5-c′]bis([1,2,5]thiadiazole) was used as an electron deficient system and electron-donating substituents include triphenylamine, thiophene, alkoxy chains, *N*, *N*-diphenyl-4-(1,2,2-triphenylvinyl)aniline 3,4-ethyl-enedioxythiophene was used as a spacer and TPE as an AIE unit to generate more red emissive unit—BPST, TB1, HLZ-BTED, BBTD suppressing twisted intramolecular charge transfer-Q8PNap	Deep tissue imaging, long-term tumor tracking monitoring due to excellent fluorescence properties	Advanced imaging-guided surgery, bimodal imaging	61

Most common approaches of AIE probes depend on either lock and key interactions or chemical and biological reactions for selective imaging and avoid nonspecific interactions with targets unlike conventional AIEgens either with charged surfaces or hydrophobic-hydrophilic interactions. Most recently, several efforts have been made to design multi-functional AIEgens having photosensitization or photothermal ability that boosted the imaging applications toward successful therapeutic strategies.

6. Conclusion, potential challenges, and future prospect

Recent AIEgens have been successfully applied for Cancer diagnosis, metastasis evaluation and possible drug screening due to the consistent improvement of optical properties, targeted therapy, and negligible toxicity.[46,47] Since its discovery AIE probes have evolved in last two decades but there are several challenges to be taken care of so that it reaches its full potential of theranostic applications. First and foremost is biocompatibility of most of the probes that needs the most attention and improvement. A clear mechanism must be established for biodegradability and as well as for fast clearance from the body for clinical applications. Several AIE examples exist that are useful in therapy, imaging and diagnosis. Clearly the multifunctional approach helps to achieve these difficult multidirectional aspects to generate positive synergistic effects. Lastly to be useful for better practical uses a standardized protocol needs to be established so that it can be transferred from lab to the clinics.

AIE based NPs have outperformed organic NPs due to their higher imaging contrast ratios and deeper penetration depth.[48] Successful skeleton tuning resulted in applications of multi-photon, NIR-II fluorescence imaging, photoacoustic imaging and even after-glow imaging.[49] Inclusion of two-photon excitation and chemiexcitation within AIE NPs offered efficient and precise image-guided PDT at deep tumor tissues in living systems. Thus, fine-tune of molecular motion in the solid state has offered smart AIE materials to be incorporated in nanoparticle systems and used in imaging, sensing, surgery and several other theranostic applications. For AIE NPs, the most common shortcoming is same as the normal nanoparticle formation which is large-scale production that remains to be addressed. One of the recent efforts described a batch dependent variation for AIE dots with a low polydispersity index and relatively higher yields using an automated microfluidic system. Uniform AIE dots were achieved using a jet mixer system and the uniformity is modulated by flow rates, solvents ratios and Reynolds number.[50]

Compared to AIE NPs, AIE dots are relatively young and serves as a powerful nanolight. Future design of AIE dots and NPs should focus on ultrabright emission in the Near IR-II region for live imaging. Structural diversity of NIR-II AIEgens is limited to a D-A-D type of skeleton and therefore, molecular engineering seeks further attention to construct diverse AIEgens.[51] Photophysical properties of most of the reported NIR-II AIEgens are below 1100 nm and the relative quantum yield is very low at the NIR-II region. Future designs should focus on higher fluorescence quantum yield at 1000–1500 nm for its potential utility at the clinics and better quality.[52] The future perspectives should also aim for the morphological controls of such NPs and dots. Till date image guided therapeutic functions were achieved using AIE dots and NPs. A multifunctional approach as discussed before can lead to future design that may include therapeutic modalities and as well as the diagnosis along with imaging would be more exciting and useful for translational studies. Although more systematic in vivo studies and worldwide collaborations are still required to close the gap for AIE materials to reach the clinics from labs. We sincerely hope that more efforts on complementary multimodal imaging and therapy will be developed in near future and AIEgens will open new windows in translational and clinical research.

Acknowledgments

This research received no external funding. SM acknowledges Dr. Omid Veiseh for continuous postdoctoral support. SC acknowledges Prof Surinder K Batra and Prof. Maneesh Jain as well as UNMC for support. SRC acknowledges his mentor Dr. Heather R Lucas at VCU for supporting his postdoctoral research.

Conflict of interest

There is no conflict of interest from any of the authors.

References

1. Bray F, Ferlay J, Soerjomataram I, Siegel RL, Torre LA, Jemal A. Global cancer statistics 2018: GLOBOCAN estimates of incidence and mortality worldwide for 36 cancers in 185 countries. *CA Cancer J Clin*. 2018;68:394–424.
2. Ferlay J, Soerjomataram I, Dikshit R, et al. Cancer incidence and mortality worldwide: sources, methods and major patterns in GLOBOCAN 2012. *Int J Cancer*. 2015;136: E359–E386.
3. DeVita Jr VT. The evolution of therapeutic research in cancer. *N Engl J Med*. 1978;298:907–910.
4. Hanahan D, Weinberg RA. Hallmarks of cancer: the next generation. *Cell*. 2011;144: 646–674.
5. Patra JK, Das G, Fraceto LF, et al. Nano based drug delivery systems: recent developments and future prospects. *J Nanobiotechnol*. 2018;16:71.

6. Chowdhury SR, Mukherjee S, Das S, Patra CR, Iyer PK. Multifunctional (3-in-1) cancer theranostics applications of hydroxyquinoline-appended polyfluorene nanoparticles. *Chem Sci*. 2017;8:7566–7575.
7. Meka RR, Mukherjee S, Patra CR, Chaudhuri A. Shikimoyl-ligand decorated gold nanoparticles for use in ex vivo engineered dendritic cell based DNA vaccination. *Nanoscale*. 2019;11:7931–7943.
8. Mukherjee S, Kotcherlakota R, Haque S, et al. Improved delivery of doxorubicin using rationally designed PEGylated platinum nanoparticles for the treatment of melanoma. *Mater Sci Eng*. 2020;108:110375.
9. Muthuraj B, Chowdhury SR, Mukherjee S, Patra CR, Iyer PK. Aggregation deaggregation influenced selective and sensitive detection of Cu2+ and ATP by histidine functionalized water-soluble fluorescent perylene diimide under physiological conditions and in living cells. *RSC Adv*. 2015;5:28211–28218.
10. Mukherjee S, Kotcherlakota R, Haque S, et al. Silver Prussian blue analogue nanoparticles: rationally designed advanced nanomedicine for multifunctional biomedical applications. *ACS Biomater Sci Eng*. 2020;6:690–704.
11. Zhuang Y, Shang C, Lou X, Xia F. Construction of AIEgens-based bioprobe with two fluorescent signals for enhanced monitor of extracellular and intracellular telomerase activity. *Anal Chem*. 2017;89:2073–2079.
12. Dutta A, Goswami U, Chattopadhyay A. Probing cancer cells through intracellular aggregation-induced emission kinetic rate of copper nanoclusters. *ACS Appl Mater Interfaces*. 2018;10:19459–19472.
13. Zhu C, Kwok RTK, Jacky W, Tang BZ. Aggregation-induced emission: a trailblazing journey to the field of biomedicine. *ACS Appl Bio Mater*. 2018;1:1768–1786.
14. Zhang CJ, Hu Q, Feng G, et al. Image-guided combination chemotherapy and photodynamic therapy using a mitochondria-targeted molecular probe with aggregation-induced emission characteristics. *Chem Sci*. 2015;6:4580–4586.
15. Zhang G, Hu F, Zhang D. Manipulation of the aggregation and deaggregation of tetraphenylethylene and silole fluorophores by amphiphiles: emission modulation and sensing applications. *Langmuir*. 2015;31:4593–4604.
16. You X, Li L, Li X, Ma H, Zhang G, Zhang D. A new tetraphenylethylene-derived fluorescent probe for nitroreductase detection and hypoxic-tumor-cell imaging. *Chem Asian J*. 2016;11:2918–2923.
17. Chen D, Wang H, Dong L, et al. The fluorescent bioprobe with aggregation-induced emission features for monitoring to carbon dioxide generation rate in single living cell and early identification of cancer cells. *Biomaterials*. 2016;103:67–74.
18. Behnke T, Mathejczyk JE, Brehm R, et al. Target-specific nanoparticles containing a broad band emissive NIR dye for the sensitive detection and characterization of tumor development. *Biomaterials*. 2013;34:160–170.
19. Bai H, Liu Z, Zhang T, et al. Multifunctional supramolecular assemblies with aggregation-induced emission (AIE) for cell line identification, cell contamination evaluation, and cancer cell discrimination. *ACS Nano*. 2020;14:7552–7563.
20. Feng G, Liu B. Aggregation-induced emission (AIE) dots: emerging theranostic nanolights. *Acc Chem Res*. 2018;51:1404–1414.
21. Feng G, Liu J, Zhang CJ, Liu B. Artemisinin and AIEgen conjugate for mitochondria-targeted and image-guided chemo- and photodynamic cancer cell ablation. *ACS Appl Mater Interfaces*. 2018;10:11546–11553.
22. Ding D, Li K, Liu B, Tang BZ. Bioprobes based on AIE fluorogens. *Acc Chem Res*. 2013;46:2441–2453.
23. Hu Q, Gao M, Feng G, Chen X, Liu B. A cell apoptosis probe based on fluorogen with aggregation induced emission characteristics. *ACS Appl Mater Interfaces*. 2015;7:4875–4882.

24. Shi HB, Kwok RTK, Liu JZ, Xing BG, Tang BZ, Liu B. Real-time monitoring of cell apoptosis and drug screening using fluorescent light-up probe with aggregation-induced emission characteristics. *J Am Chem Soc*. 2012;134:17972–17981.
25. Zhan C, Zhang G, Zhang D. Zincke's salt-substituted tetraphenylethylenes for fluorometric turn-on detection of glutathione and fluorescence imaging of cancer cells. *ACS Appl Mater Interfaces*. 2018;10:12141–12149.
26. Zhang P, Zhao Z, Li C, et al. Aptamer-decorated self-assembled aggregation-induced emission organic dots for cancer cell targeting and imaging. *Anal Chem*. 2018;90: 1063–1067.
27. Che W, Zhang L, Li Y, et al. Ultrafast and noninvasive long-term bioimaging with highly stable red aggregation-induced emission nanoparticles. *Anal Chem*. 2019;91: 3467–3474.
28. Nicol A, Qin W, Kwok RT, et al. Functionalized AIE nanoparticles with efficient deep-red emission, mitochondrial specificity, cancer cell selectivity and multiphoton susceptibility. *Chem Sci*. 2017;8:4634–4643.
29. Tian H, Li D, Tang X, et al. Efficient red luminogen with aggregation-induced emission for in vivo three-photon brain vascular imaging. *Mater Chem Front*. 2020;4:1634–1642.
30. Li K, Qin W, Ding D, et al. Photostable fluorescent organic dots with aggregation-induced emission (AIE dots) for noninvasive long-term cell tracing. *Sci Rep*. 2013;3:1150.
31. Chen Y, Li M, Hong Y, Lam JW, Zheng Q, Tang BZ. Dual-modal MRI contrast agent with aggregation-induced emission characteristic for liver specific imaging with long circulation lifetime. *ACS Appl Mater Interfaces*. 2014;6:10783–10791.
32. Ma B. Two-photon AIE probe conjugated theranostic nanoparticles for tumor bioimaging and pH-sensitive drug delivery. *Nano Res*. 2019;12:1703–1712.
33. Han X, Liu DE, Wang T, et al. Aggregation-induced-emissive molecule incorporated into polymeric nanoparticulate as FRET donor for observing doxorubicin delivery. *ACS Appl Mater Interfaces*. 2015;7:23760–23766.
34. Dai Y, He F, Ji H, Zhao X, Misal S, Qi Z. Dual-functional NIR AIEgens for high-fidelity imaging of lysosomes in cells and photodynamic therapy. *ACS Sens*. 2020;5:225–233.
35. Wu WB, Feng GX, Xu SD, Liu B. A photostable far-red/near-infrared conjugated polymer photosensitizer with aggregation-induced emission for image-guided cancer cell ablation. *Macromolecules*. 2016;49:5017–5025.
36. Jin G, He R, Liu Q, et al. Near-infrared light-regulated cancer theranostic nanoplatform based on aggregation-induced emission luminogen encapsulated upconversion nanoparticles. *Theranostics*. 2019;9:246–264.
37. Gao H, Bao P, Dai S, et al. Far-red/near-infrared emissive (1,3-dimethyl)barbituric acid-based AIEgens for high-contrast detection of metastatic tumors in the lung. *Chem Asian J*. 2019;14:871–876.
38. Gao Y, Zheng QC, Xu S, Yuan Y, Cheng X, Jiang S. Theranostic nanodots with aggregation-induced emission characteristic for targeted and image-guided photodynamic therapy of hepatocellular carcinoma. *Theranostics*. 2019;9:1264–1279.
39. Ni X, Zhang X, Duan X, Zheng HL, Xue XS, Ding D. Near-infrared afterglow luminescent aggregation-induced emission dots with ultrahigh tumor-to-liver signal ratio for promoted image-guided cancer surgery. *Nano Lett*. 2019;19:318–330.
40. Cai X, Liu B. Aggregation-induced emission: recent advances in materials and biomedical applications. *Angew Chem Int Ed Engl*. 2020;59:9868–9886.
41. Feng G, Liu B. Multifunctional AIEgens for future theranostics. *Small*. 2016;12: 6528–6535.
42. Zhang C, Zhao Y, Li D, et al. Aggregation-induced emission (AIE)-active molecules bearing singlet oxygen generation activities: the tunable singlet-triplet energy gap matters. *Chem Commun (Camb)*. 2019;55:1450–1453.

43. Xu S, Yuan Y, Cai X, et al. Tuning the singlet-triplet energy gap: a unique approach to efficient photosensitizers with aggregation-induced emission (AIE) characteristics. *Chem Sci*. 2015;6:5824–5830.
44. Shen Y, Shuhendler AJ, Ye D, Xu JJ, Chen HY. Two-photon excitation nanoparticles for photodynamic therapy. *Chem Soc Rev*. 2016;45:6725–6741.
45. Cai XL, Hu F, Feng GX, Kwok RTK, Liu B, Tang BZ. Organic mitoprobes based on fluorogens with aggregation-induced emission. *Isr J Chem*. 2018;58:860–873.
46. Gu XG, Tang B. No UV irradiation needed! chemiexcited AIE dots for cancer theranostics. *Chem*. 2017;3:922–924.
47. Gu X, Kwok RTK, Lam JWY, Tang BZ. AIEgens for biological process monitoring and disease theranostics. *Biomaterials*. 2017;146:115–135.
48. Li D, Qin W, Xu B, Qian J, Tang BZ. AIE nanoparticles with high stimulated emission depletion efficiency and photobleaching resistance for long-term super-resolution bioimaging. *Adv Mater*. 2017;29:1703643.
49. Sheng Z, Guo B, Hu D, et al. Bright aggregation-induced-emission dots for targeted synergetic NIR-II fluorescence and NIR-I photoacoustic imaging of orthotopic brain tumors. *Adv Mater*. 2018;30:e1800766.
50. Valencia PM, Farokhzad OC, Karnik R, Langer R. Microfluidic technologies for accelerating the clinical translation of nanoparticles. *Nat Nanotechnol*. 2012;7:623–629.
51. Li Y, Hu D, Sheng Z, et al. Self-assembled AIEgen nanoparticles for multiscale NIR-II vascular imaging. *Biomaterials*. 2020;264:120365.
52. Xu W, Wang D, Tang BZ. NIR-II AIEgens: a win-win integration towards bioapplications. *Angew Chem Int Ed Engl*. 2020;133:7552–7563.

CHAPTER THREE

AIE material for photodynamic therapy

Vishal Saini and V. Venkatesh*
Department of Chemistry, Indian Institute of Technology Roorkee, Roorkee, Uttarakhand, India
*Corresponding author: e-mail address: venkatesh.v@cy.iitr.ac.in

Contents

Abstract

Photodynamic therapy (PDT) is emerging as an excellent strategy to treat different types of cancers. The advantages of using PDT over other cancer treatment modalities are owing to its non-invasive nature, spatiotemporal precession, controllable photoactivity, and least side effects. The photosensitization ability of traditional photosensitizers (PSs) are severely curtailed by aggregation-induced quenching (ACQ). On the contrary, aggregation induced emission (AIE) molecules/fluorogens (AIEgens) show enhanced fluorescence emission and high reactive oxygen species (ROS)/singlet oxygen ($^1O_2$) production capability in the aggregated state. These unique characteristics of AIEgens make them potential AIE-PSs for fluorescence/luminescence image-guided combination PDT. In this chapter, we discussed the strategies that are developed to synthesize small molecule-based AIE-PSs, metal complex-based AIE-PSs, and AIE-PSs with two-photon absorbance (TPA) properties, polymer-based AIE-PSs, and nanoparticles based AIE-PSs for PDT. We have also discussed the rational design of targeting peptide

Progress in Molecular Biology and Translational Science, Volume 185
ISSN 1877-1173
https://doi.org/10.1016/bs.pmbts.2021.06.010

conjugated AIE-PSs to selective target cancer cells over normal cells. Furthermore, recent findings on nanoparticle-based combination AIE-PSs are also discussed, where the combination AIE-PSs show synergistically improved anticancer activity and overcome the drug resistance. Finally, we shed light on the recent development, ongoing challenges, and future directions for designing better AIE-PS for PDT.

Graphical abstract

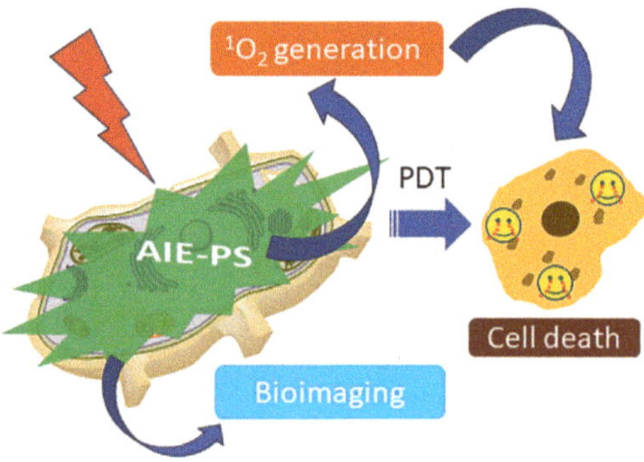

1. Introduction

Cancer is the deadliest and devastating disease worldwide. Besides other conventional cancer treatments, such as chemotherapy, radiotherapy, and surgical treatments, photodynamic therapy (PDT) is emerging as a promising cancer treatment strategy with high spatiotemporal precision, non-invasive nature, and least side effects.[1] PDT relies upon photosensitizer (PS), light and oxygen molecule for its activity. The photoactivation of PS with a particular wavelength of light excites PS to higher singlet S_n excited state, and then it can relax back to S_1 via internal conversion. From S_1 state, the PS can decay back to S_0 state or undergo intersystem crossing (ISC) to populate PS in long lived triplet T_1 state. The PS in the T_1 state can transfer its energy via type I (free radical generation/electron transfer) or type II (production of singlet oxygen ($^1O_2$)/reactive oxygen species (ROS)) process. The energy released from the excited T_1 state of PS can produce ROS/$^1O_2$, responsible for their PDT activity. There are also many other advantages of using PDT in cancer treatment. Most PSs exhibit bright fluorescence/luminescence emission that can be useful to monitor the organelle specificity, PDT activity and possible to moderate the PDT activity using light activatable strategy.[2]

Many PSs developed for PDT treatment, out of which photofrin, 5-aminolevulinic acid, and methyl aminolevulinate are the three food and drug administration (FDA) approved PDT agents available to treat certain types of cancers.[3] Design of PSs with selective tumor targeting ability, low dark toxicity, near-infrared (NIR) activation and efficient ROS/$^1O_2$ generation is highly desirable, whereas currently developed PSs are far from the ideal PSs. The well-studied PSs, such as porphyrin, chlorin, phthalocyanine, 4,4-difluoro-4-bora-3a,4a-diaza-s-indacene (BODIPY) and their analogs, can undergo π-π stacking in the aggregated state due to their rigid planar structure (Fig. 1A). Such stacking interactions may initiate non-radiative pathways with a simultaneous reduction in ROS/$^1O_2$ production, as result of ACQ (Fig. 1C). In 2001, Prof. B. Z. Tang and co-workers explored an interesting phenomenon called AIE.[4] According to the AIE phenomenon, the fluorophores are poorly emissive/non-emissive in the dissolved state, whereas they emit strongly in the aggregated state. Compared to traditional PSs, the propeller-shaped conformation of AIE-PSs suppresses π-π stacking in the aggregated state owing to restricted intramolecular motions (RIMs) (Fig. 1B), thereby reducing non-radiative decay. Besides, AIEgens with

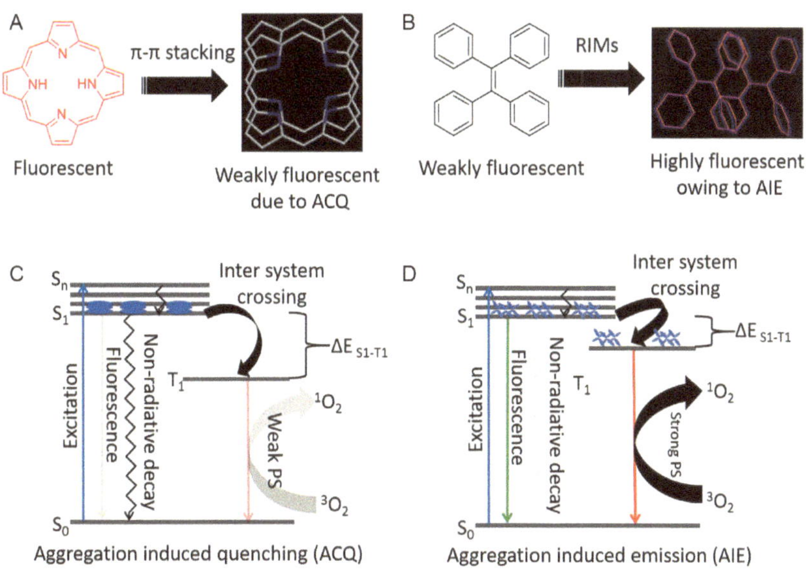

Fig. 1 (A) and (C) Represent ACQ of porphyrin-based PS in its aggregated state and its corresponding electronic transitions. (B) and (D) are AIE of tetraphenylethene (TPE) in its aggregated state and its corresponding electronic transitions.

D-π-A skeleton reduce the energy between first singlet (S_1) and first triplet (T_1) excited states ($\Delta E_{S1\text{-}T1}$) and populate T_1 state that can potentiate the photosensitization of PS for PDT (Fig. 1D).

AIEgens with large stokes shift, excellent photostability, good biocompatibility and better tolerance toward high concentration attracted considerable interest in the area of bioimaging, biosensing and PDT.[5] Recently, small molecules, metal complexes, polymers and nanoparticles with AIE characteristics are emerging as a turn-on fluorescent probe for detection and efficient PSs for PDT.[6,7] Thus, the AIEgens endowed with light-up fluorescence and excellent PS capability can act as a promising agent for image-guided PDT for dermatology, ophthalmology, infectious bacterial eradication and cancer cell ablation. This chapter provides an overview of the design strategies, organelle targeting abilities, light activation (one photon/two-photon), and recent updates of the developed AIEgens as PSs in PDT. Finally, we discuss the limitations, challenges, and future scopes for AIEgens in the area of image-guided PDT.

2. Small molecule-based organelle targeting AIEgens for PDT

Small molecule-based organelle targeting AIE-PSs are in high demand for designing effective fluorescence image-guided photodynamic therapy. Targeting moieties are deployed to design AIEgens with specific organelles targeting abilities, whereas introducing targeting groups on AIEgens is a tedious process and may lead to a change in the properties of AIEgens. Therefore, the rational design of AIEgens with inherent targeting abilities is highly challenging and desirable in the area of organelle targeting AIEgens for PDT.

2.1 Mitochondria targeting AIEgens

Mitochondria, the energy source of cells, crucial for deciding the life or death of cells. Targeting mitochondria in cancer cells is an important strategy to kill cancer cells selectively. Many AIEgens have been developed to target cancer cells selectively. Most of the reported AIEgens are mitochondria specific and exert potent PDT activity (Fig. 2A). Prof. B. Z. Tang's group is extensively involved in exploring the applications of AIEgens for bioimaging, biosensing, and therapy. Recently, they reported the synthesis and cancer theranostic applications of TTPy and MeTTPy ligands (Fig. 2B).[8] The ligand design consists of D-Π-A skeleton; it facilitates intramolecular

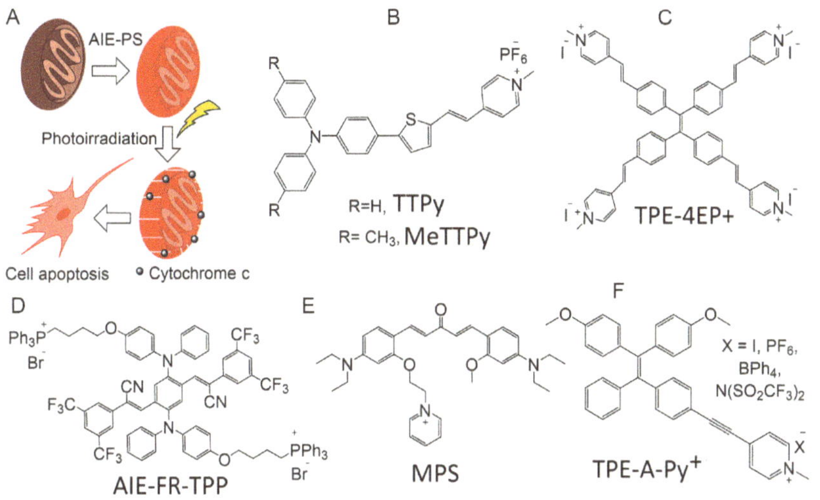

Fig. 2 (A) Schematic representation of AIE-PS binding to mitochondria, followed by turn-on fluorescence, and photo-irradiation initiates the release of cyt *c* and further leads to cell apoptosis. (B–F) Chemical structures of AIE-PSs containing pyridinium/triphenylphosphine cation moiety in their molecular structure target mitochondria for their PDT activity.

charge transfer and exhibits AIE in different solvent systems. Cell imaging studies carried on HeLa cells clearly depict that both TTPy and MeTTPy are localizing in cells mitochondria. The positively charged pyridinium ion present in the ligands might be attracted toward negatively charged mitochondrial transmembrane potential, notably higher mitochondrial potential of cancer cells (60 mV) compared with normal cells pave the way to target cancer cells over normal cells selectively. To check this hypothesis, the authors incubated the ligands with normal and cancer cells for 20 min and imaged with the help of confocal laser scanning microscopy (CLSM). Bright emission was observed in cancer cells, whereas weak emission was observed with normal cells indicating the selective targeting abilities of AIEgens toward cancer cells. The ligands ROS generating ability was tested after photoactivation with white light using 2′,7′-dichlorodihydrofluorescein diacetate (DCF-DA) assay. The ROS quantum yield (90.7% for MeTTPy) of the ligands were compared with known PSs (Ce6 and Rose Bengal). It was noted that the ROS quantum yield of ligands is much higher than previously reported PSs. 3-(4,5-Dimethylthiazol-2-yl)-2,5-diphenyltetrazolium bromide (MTT) assay shows least toxicity of the ligands under dark condition suggesting their biocompatibility; upon white light exposure, cell viability is drastically reduced, particularly in cancer cells.

In another study, they designed AIEgens with four pyridinium salt TPE-4EP+ (Fig. 2C) with aggregation-induced emission features.[9] The rationale for designing AIEgens with high positive charge might facilitate the electrostatic interaction of TPE-4EP+ with high negative mitochondria membrane potential present in cancer cells. As expected, the CLSM images of TPE-4EP+ treated HeLa cells showed bright emission and high specificity for mitochondria. It was interesting to note that upon white light irradiation resulted in movement of TPE-4EP+ to the nucleus. Irradiation of TPE-4EP+ might depolarize mitochondrial membrane potential and increases nuclear membrane permeability. Thus, detached TPE-4EP+ bind with DNA present in cell nuclease and fluoresce. Similarly, H. Gao and co-workers[10] developed a multifunctional AIEgen probe AIE-FR endowed with mitochondria targeting triphenylphosphonium salt (TPP) (Fig. 2D). AIE-FR-TPP probe selectively stains mitochondria even at low concentration, upon white light irradiation, it produced ROS with high quantum yield. They have monitored the PDT activity of this probe in real-time on live zebrafish embryos; photo-induced ROS generation leads to the collapse of the mitochondrial membrane and concomitant morphological changes were observed.

C. Yang *et al.* designed a PS, MPS from dibenzylideneacetone derivative containing mitochondria targeting pyridinium cation (Fig. 2E).[11] MPS exhibits bright red emission in the aggregated state (non-polar solvents) and weak emission in the dissolved state (aqueous medium). Co-localization experiment with MitoTracker Green FM confirmed its accumulation in mitochondria; further irradiation with white light generates ROS. The ROS produced may cause mitochondrial membrane permeabilization (MMP) and cause MPS's translocation from less polar mitochondria to highly polar cytosol. This polarity difference causes the fluorescence quenching and it ceases photosensitization ability in the cytosol. Such a self-traceable MPS can find immense application in traceable PDT. D. Zhang and co-workers demonstrated the effect of anion and the presence of alkyne linker in pyridinium substituted TPE for their efficient photodynamic activity.[12] They have synthesized pyridinium substituted TPE with alkyne linker TPE-A-Py+ and without alkyne linker TPE-Py+ and varied the counter anion such as I^-, PF_6, $B(ph_4)^-$, $N(SO_2CF_3)^{2-}$ (Fig. 2F). Their comprehensive study concludes that the presence of the alkyne linker reduces the energy difference between S_1 and T_1 excited states ($\Delta E_{S1\text{-}T1}$) and the presence of heavy I^- anion facilitate ISC, resulting in the potential PDT activity of TPE-A-Py+ against Gram-positive and Gram-negative bacteria in the presence of white light.

2.2 Lysosome targeting AIEgens

Lysosome plays a significant role in cellular processes such as cell signaling, cholesterol homeostasis and activating cellular apoptosis, to name a few. In this regard, efforts have been made to target lysosome in cancer cells. Recently, Y. Dai *et al.* reported the synthesis and specific lysosome targeting abilities of AIEgens.[13] They have designed three new AIEgens (MPAT, MPAN and MPAA) containing morpholine moieties in their structural design (Fig. 3A–C). The presence of a basic morpholine group is crucial for targeting the acidic environment of the lysosome. They have also demonstrated that one of the AIEgens, MPAT, can able to dynamically track the lysosome in live cells and zebrafish model. The PS characteristic of MPAT was explored after irradiating it with green light, the production of ROS/$^1O_2$ was confirmed from DCF-DA assay and the phototoxicity was evaluated against HeLa cells.

In another recent study by J. Zhuang *et al.* synthesized two AIEgens such as MTMM and PTMM (Fig. 3D) from morpholinyl and piperidyl functionalized *p*-bis(2,2-dicyanovinyl)benzene.[14] The lipophilicity, pK_a and basic morpholine group present in the AIEgens facilitate their localization into

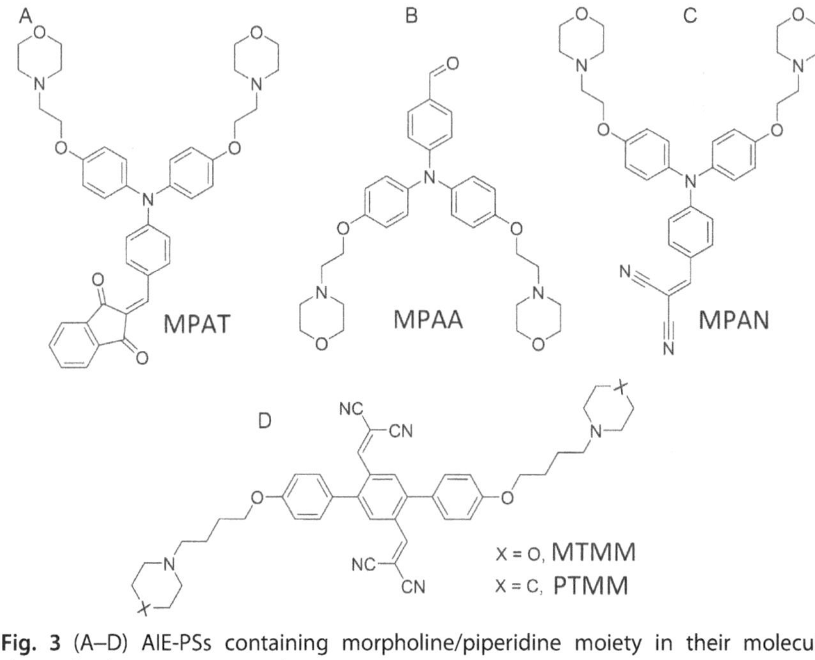

Fig. 3 (A–D) AIE-PSs containing morpholine/piperidine moiety in their molecular design for lysosome targeted PDT.

lysosome and it was evident from the co-localization experiment performed with LysoTracker Red. Upon white light irradiation, both MTMM and PTMM produces $^1O_2$ and ablate cancer cells. The photobleaching/degradation of 9,10-anthracenediyl-bis(methylene)dimalonic acid (ABDA) with AIEgens after white light irradiation signifies ROS/$^1O_2$ generation. Besides, these AIEgens discriminate Gram-positive bacteria from Gram-negative bacteria. Treatment of AIEgens with *Staphylococcus aureus* and *Escherichia coli* resulted in bright yellow emission emerging from *S. aureus*, whereas no notable emission signals were observed from *E. coli*. Their selective localization in Gram-positive bacteria can selectively inhibit the colony-forming unit (CFU) of *S. aureus* under white light irradiation; under similar light irradiation condition, CFU of *E. coli* is unaffected.

2.3 Other organelles and multi-organelles targeting AIEgens

Recently, B. Z. Tang and co-workers have designed a double charged pyridinium salt functionalized tetraphenylethene TPE-MEM (Fig. 4A) with AIE feature, to target cell membrane selectively.[15] Interestingly, they have noticed that TPE-MEM produces reactive oxygen species (ROS) at room light irradiation with simultaneous staining of cell membrane. The reactive oxygen produced can oxidize cholesterol, phospholipids and membrane proteins that lead to the damage plasma membrane and cause cell necrosis.

AIEgens targeting lipid droplet (LD) were developed from triphenylamine (TPA) and thiophene building blocks.[16] The high solid state fluorescence quantum yields of TPA-thiophene conjugates are presumably due to their intrinsic AIE behavior. It was noted that they selectively stain LD with better image contrast and photostability than the commercially available LD imaging probe. In addition to specific staining of LD, one of the conjugates, TTNIR (Fig. 4B), produces ROS upon irradiation with white light enabling it to act as an efficient photosensitizer for photodynamic therapy.

W. Xu *et al.* recently come up with a new strategy to target multiple organelles of cells simultaneously.[17] In this context, they have designed three homologous AIEgens such as TFPy to target mitochondria, TFVP to target cell membrane and TPE-TFPy to target lysosome (Fig. 4C–E). All the three AIEgens can produce ROS after irradiation with white light, a combination of all three AIEgens produces ROS derived from different organelles and exhibit potent phototoxicity against in vitro and in vivo tumor models even at very low concentrations (2.5 μM). However, at similar concentrations the

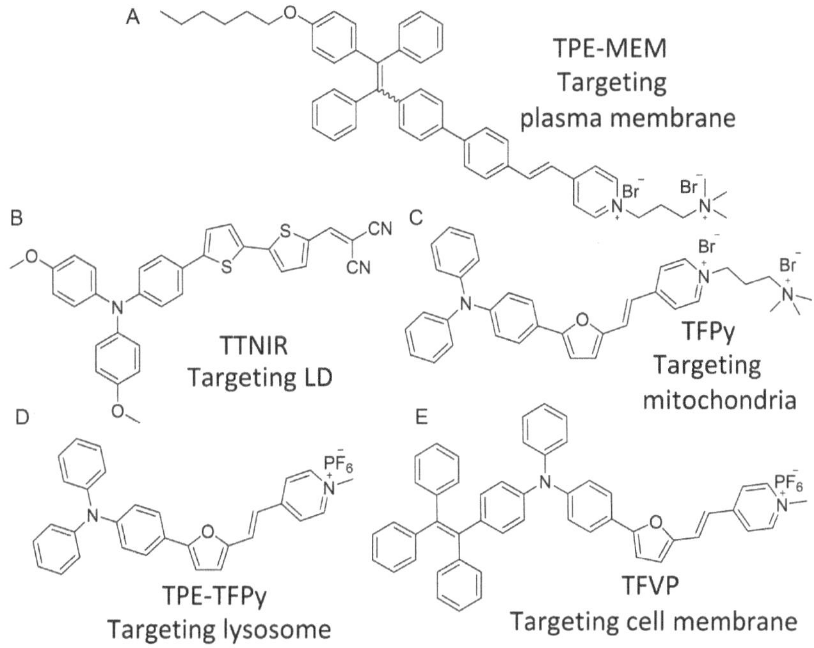

Fig. 4 Chemical structures of (A) AIE-PS having double charged pyridinium ion with balanced hydrophilicity and hydrophobicity to target plasma membrane. (B) Lipid droplet targeting AIE-PSs. (C–E) Three-in-one AIEgens to target mitochondria, lysosome and cell membrane, respectively.

phototoxicity of individual AIEgens is negligible. Superior therapeutic efficacies of three-in-one AIEgen is due to three-pronged approach to attack the cancer cells with $1+1+1>3$ synergistic PDT.

3. Small-molecule based AIEgens for two-photon PDT

In the above discussed examples, activation of AIEgens as PSs for PDT requires UV/Visible light for their activity. However, these higher energy radiations are poorly penetrating inside tissues and potentially harmful to healthy cells. In this regard, two-photon absorption AIEgens can act as an efficient photosensitizer and their activation requires lesser energy, higher penetration near-infrared (NIR) radiation (Fig. 5A).

B. Z. Tang and co-workers developed AIEgens for two-photon PDT (TP-PDT).[18] One of the synthesized molecules, IQ-TPA (Fig. 5B) possess high two-photon cross-section ($\delta_{IQ\text{-}TPA}=213$ GM), high specificity to

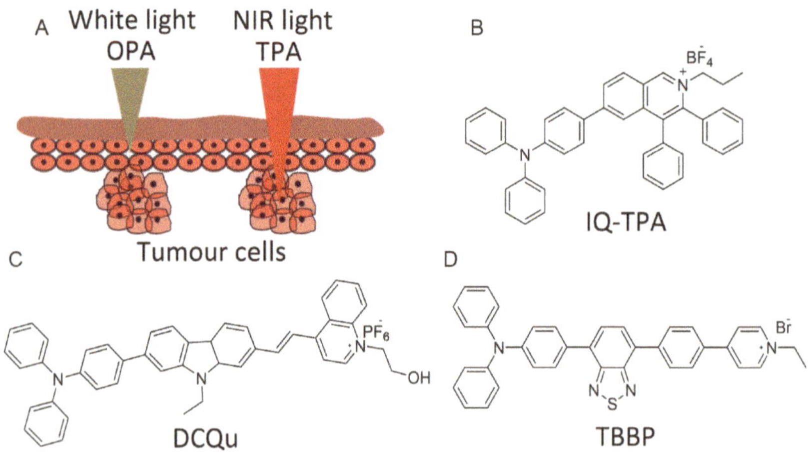

Fig. 5 (A) Schematic representation compares the merits of two-photon excitation process such as lower excitation wavelength, high light penetration depth and focused spot excitation over the one-photon excitation process. (B–D) AIE-PSs with D-π-A skeleton for two-photon active PDT.

mitochondria and exhibits AIE feature. IQ-TPA act as a potential photosensitizer for image-guided TP-PDT. The ability of this molecule to generate intracellular ROS was monitored using DCF-DA after irradiation with 900 nm light. As IQ-TPA was found to localize in the mitochondria, they have further studied the structural changes of mitochondria with CLSM after different scans of two-photon irradiation. The fragmented mitochondrial structure was monitored after 100 scans of two-photon irradiation at 900 nm, supports the mitochondria-mediated cell death.

The same group has recently designed AIE-PSs with multiple characteristics.[19] They have designed carbazole functionalized pyridinium moieties (CPy) with different D-A strength. One of the designed compounds, DCQu (Fig. 5C) shows the least energy gap between first singlet and triplet excited state ($\Delta E_{st} = 0.445$ eV). DCQu shows multifaceted properties such as efficient $^1O_2$ production, selective accumulation inside cancer cells, mitochondria localization, activation with NIR excitation and bright NIR emission makes it as a promising agent for image guided TP-PDT.

Recently, Y. Wang and co-workers designed a photosensitizer TBBP endowed with AIE behavior for TP-PDT.[20] The structural design of TBBP consists of the D-π-A skeleton, where triphenylamine act as a donor, pyridinium ion act as acceptor moiety and phenyl benzothiadiazole group act as a π conjugate system (Fig. 5D). The presence of electron-withdrawing pyridinium ion was employed to target cellular mitochondria. The two-photon cross-section of TBBP is 198 GM, two-photon bioimaging of live cells with TBBP at 800 nm excitation shows the selective staining of mitochondria. The authors performed two photon mitochondrial imaging studies on mice models to study deep tissue imaging. The experimental results showed TBBP accumulates predominantly on liver tissues and tissue imaging is possible up to 150 μm depth.

4. Metal complexes as AIEgens for PDT

Transition metal complexes having multiple interesting characteristics such as impressive photophysical properties, high triplet state lifetime and large two-photon cross-section are potential TP-PDT candidates. The first report of metal complexes targeting mitochondria with AIE feature for TP-PDT was reported by H. Chao and co-workers.[21] They have developed cyclometallated iridium complexes Ir1-Ir3 (Fig. 6A) as mitochondria targeting AIE-PSs for TP-PDT. The TPA cross-section of Ir1 was found to be 214 GM, which is better than any of the reported bioactive organometallic probe. The abilities of theses PSs to produce $^1O_2$ upon irradiation

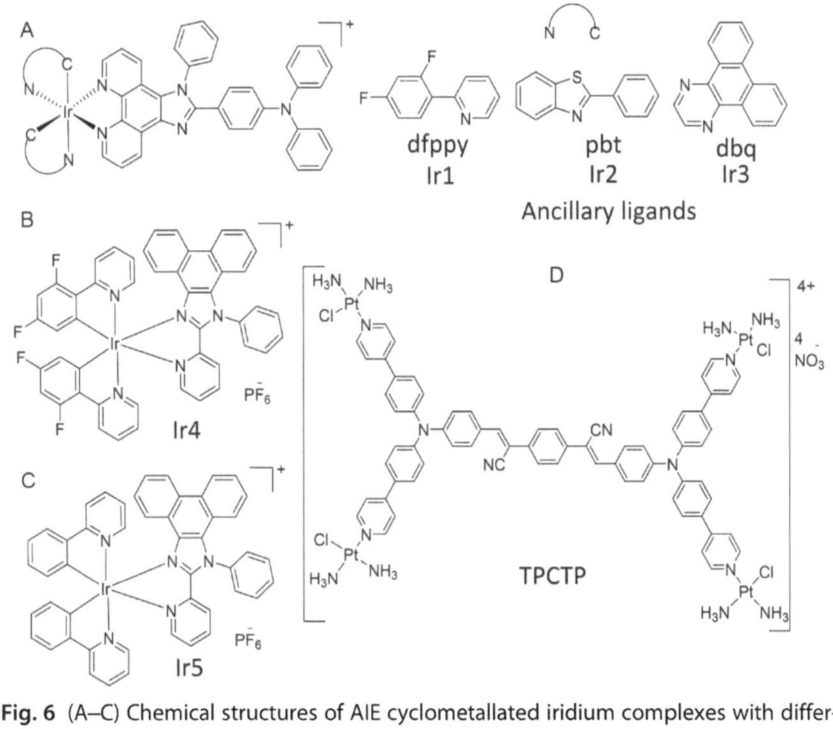

Fig. 6 (A–C) Chemical structures of AIE cyclometallated iridium complexes with different ancillary ligands and (D) molecular structure of AIE platinum complex TPCTP for PDT.

were determined using $^1O_2$ phosphorescence intensities. Ir1 exhibits potent phototoxicity against HeLa cells in one and two-photon activation, notably TPA of Ir1 on 3D multicellular tumor spheroids showed high lethality due to its deeper penetration depth upon TPA. Whereas Ir2 and Ir3 are least toxic upon TPA, indicating their low TPA cross-sections and therefore, poor ROS generating abilities. In the follow-up study, they have designed mitochondria targeting Ir4 and Ir5 (Fig. 6B and C) and studied their TP-PDT activities.[22] Ir4 and Ir5 form nanoaggregates in water and it was characterized with the help of transmission electron microscope (TEM). Light irradiation of Ir4 and Ir5 produces $^1O_2$, which was trapped with help of 2,2′,6,6′-tetramethylpiperidine (TEMP) spin trap and characterized using electron paramagnetic resonance (EPR) spectroscopy. The cellular uptake of these complexes was monitored by two-photon imaging facilities of CLSM. The uptake of Ir4 and Ir5 depend upon their incubation time, increasing the incubation time produce enhanced phosphorescence emission due to increased aggregation of complexes in cells. The mitochondrial damage was observed when Ir4 or Ir5 was irradiated with one or two-photon irradiation. CLSM images showed the morphology of mitochondria changes from normal threadiness to a solid granule and finally to a hollow sphere, extended irradiation causes shrinkage of cells. In the case of Ir4 and Ir5, the phototoxicity index (PI) of two-photon irradiation was higher than that of one photon irradiation, demonstrating effectiveness of TPA over OPA.

Recently, Z. He *et al.* reported the synthesis of platinum-AIEgen complex TPCTP (Fig. 6D).[23] It was prepared by the reaction of tetra-pyridyl substituted TPC (AIEgen) with cisplatin. The improved water solubility of TPCTP shows weak fluorescence emission at higher water content when compared with AIEgen TPC. The high positive charge on TPCTP bind negatively charged albumin with high affinity due to electrostatic and hydrophobic interactions. The binding of TPCTP with albumin improved the fluorescence emission and exhibited strong fluorescence emission at higher albumin concentration due to AIE. The $^1O_2$ quantum yield of TPCTP was found to be 0.75, which is the highest $^1O_2$ quantum yield that has been reported so far for any platinum-based PDT agents. The uptake studies revealed that the higher uptake was monitored against cisplatin-resistant (A549/DDP, HeLa/DDP) cells. This selectivity could be due to the different cytomembrane structures between resistant and normal cancer cells. The cytotoxicity MTT assay results show TPCTP exert potent phototoxicity against cisplatin-resistant cancer cells, which may be due to its higher uptake and can be regarded as an effective therapy for platinum-resistant tumors.

5. Targeted peptide conjugated AIEgens for PDT

F. Hu *et al.* developed the first dual-targeted AIE-PS probe TPE-red-2AP2H (Fig. 7A) for targeted bioimaging and PDT.[24] Red emissive probe TPE-red was attached with 2AP2H (IHGHHIISVG), to target hydrophilic extracellular loop EL2 (PYRDDVMSVN) and lysosome transmembrane protein LAPTM4B that are overexpressed in most of the solid tumors. The fluorescence emission of TPE-red-2AP2H is weak, but after EL2/LAPTM4B binding resulted in bright red emission due to the restriction in free rotation and interestingly their emission was much brighter under acidic conditions (pH 5.5) that is commonly observed in cancer cells. The probe not only useful for bio-imaging of transmembrane protein LAPTM4B, but also trace their movement inside the cells. PS capability of the probe was evaluated, bleaching of anthracene 9,10-dipropionic acid (ADPA) was monitored when it was irradiated with the probe at 450 nm light, indicating the formation of $^1O_2$. The probe is very selective for LAPTM4B, this was confirmed when the cells devoid of LAPTM4B was treated with TPE-red-2AP2H, light up emission and PDT activities are inhibited. Therefore, this probe is suitable for targeted imaging and targeted PDT applications. A dual-targeted activatable AIE-PS probe was designed from fluorophore with AIE characteristic, biothiols cleavable quencher groups and a targeting cyclic peptide arginine-glycine-aspartic acid (cRGD) (Fig. 7B).[25] The presence of cRGD moiety in the probe selectively target overexpressed $\alpha_v\beta_3$ integrin in cancer cells and get internalized via receptor-mediated endocytosis. In cancer cells, high levels of glutathione (GSH) are present due to aberrant metabolism, facilitating the cleavage of 2,4-dinitrobenzenesulfonyl (DNBS) group and activating the PS. The dual targeting characteristic of the probe light-up the fluorescence and improve photosensitizing ability. They have performed fluorescence imaging studies to dissect the targeting ability of the probe. For this, they have treated the probe with cancer cells that overexpress $\alpha_v\beta_3$ integrin (MDA-MB-231/U87-MG) and cancer cells that express low $\alpha_v\beta_3$ integrin (MCF-7). The red fluorescence emission was noted in the case of MDA-MB-231/U87-MG cells and its emission intensity increases with increasing incubation time. On the contrary, a weak emission was observed with MCF-7 cells. Similarly, blocking the synthesis of GSH using buthionine sulfoximine (BSO) also resulted in weak fluorescence emission signifying the probes specific targeting abilities toward overexpress $\alpha_v\beta_3$ integrin and GSH.

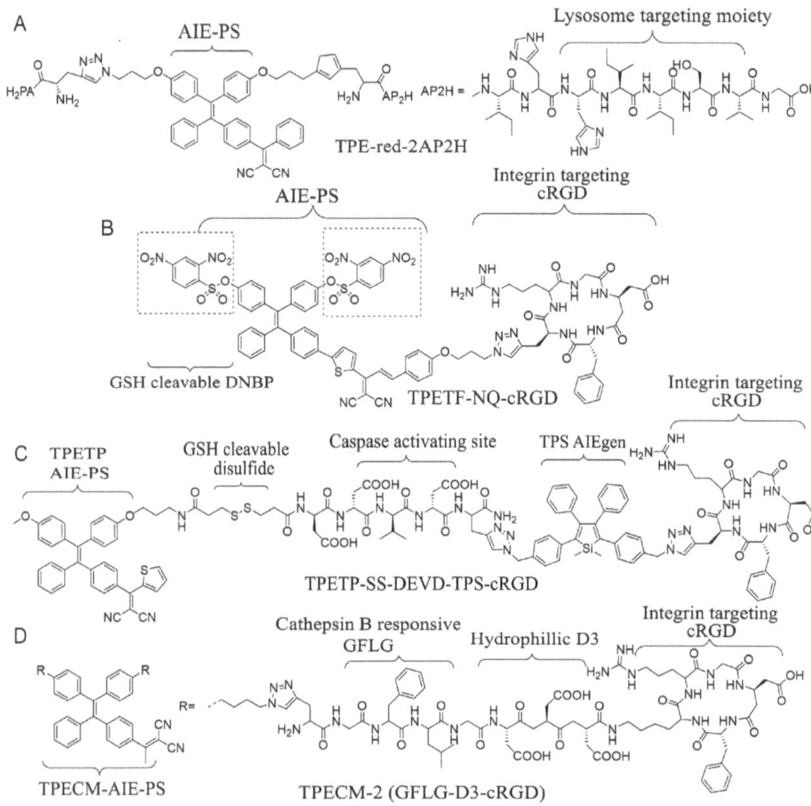

Fig. 7 Peptide conjugated AIE-PSs for targeted PDT. (A) Lysosomal transmembrane protein targeting TPE-red-2AP2H. (B) Integrin targeting and GSH activatable TPETF-NQ-cRGD. (C) Caspase activating, integrin targeting and GSH activatable TPETP-SS-DEVD-TPS-cRGD. (D) Integrin targeting and cathepsin B responsive TPECM-2 (GFLG-D3-cRGD).

AIE-PS with self-reporting the activation and therapeutic responses by dual fluorescence turn-on emission at a single excitation wavelength was developed.[26] The probe is composed of TPETP-SS-DEVD-TPS-cRGD (Fig. 7C). The presence of cRGD tripeptide selective internalizes the probe into cancer cells via receptor-mediated endocytosis. Disulfide linkage (S—S) activates the probe toward GSH and subsequently releases the red fluorescent TPEP and non-emissive apoptosis sensor. Irradiation of the probe produces ROS that activates apoptosis and apoptosis caspases enzymes cleaving the DEVD substrate and releasing green fluorescent TPS for real time monitoring the therapeutic action. In brief, the activation of AIE-PS can be monitored from its red fluorescence emission. The initiation of apoptosis will be inferred from its green fluorescence emission at the single excitation wavelength that can be beneficial for monitoring the drug activation processes and therapeutic responses.

In another studies, they have developed enzyme activatable AIEgens as a photosensitizer for photodynamic therapy and simultaneous light-up fluorescence probe.[27] The probe consists of AIEgen, cathepsin B responsive GFLG peptide, hydrophilic aspartic acid units (D3) and cRGD peptide (Fig. 7D). The presence of hydrophilic aspartic acid units, peptide chain improved the aqueous solubility of the probe. It exhibits negligible fluorescence with the least ROS generation upon irradiation in an aqueous medium, makes it to act as a biocompatible probe. Upon treatment with cancer cells, overexpressed cathepsin B triggers GFLG peptide cleavage from the probe that leads to aggregation of AIEgen and enhances the fluorescence emission and increases ROS generation after light irradiation. Cathepsin B triggered probe activation was monitored using HPLC. After cathepsin B treatment, the probe's retention time has increased due to increased hydrophobicity. The probe's selectivity was tested against different biomolecules such as GSH, cysteine, bovine serum albumin (BSA), deoxyribonucleic acid (DNA), trypsin, etc. The increased fluorescence emission was only observed after cathepsin B treatment precludes the probe's premature activation before reaching the target. The ROS generating ability of the probe after photo-irradiation was studied with help of 1,3-diphenylisobenzofuran (DPBF). The change in the absorbance of DPBF at 418 nm was monitored after irradiating the probe with DPBF, minimal reduction in the absorbance was noted. In another experiment, the probe was pretreated with cathepsin B, followed by irradiation of the probe with DPBF results in a drastic reduction in DPBF absorbance. The ability of the cathepsin B pretreated probe to generate ROS after light irradiation was further confirmed using DCF-DA assay. The probe was treated with cancer cells (MDA-MB-231 and MCF-7) and non-cancerous human cells (293T cells). The CLSM images show bright emission from high cathepsin B expressing MDA-MB-231 cells and weak emissions were observed from low cathepsin B expressing MCF-7 and 293T cells. Cathepsin B is known as a lysosomal protease, to dissect the activation site of the probe in cells, colocalization experiments were carried out with LysoTracker Green DND-26. The colocalization experiment confirmed lysosome is the target for the activated probe and the light irradiation of the cells can cause lysosomal damage, which was further ascertained from acridine orange staining. The cytotoxicity of the probe was evaluated with and without light irradiation using MTT assay.

The probe exhibits the least toxicity against all the tested cell lines in the absence of light, whereas in the presence of light it exhibits potent cytotoxicity against MDA-MB-231 cells.

6. Combination of chemotherapeutic agent functionalized AIE-PS for chemo-photodynamic therapy

F. Hu *et al.* designed a probe, TPEPY-S-MMC by integrating activatable PS and chemo-prodrug for the first time.[28] The dual prodrugs are non-toxic, electron-withdrawing acyl substituent on Mitomycin C (MMC) inhibited its cytotoxicity and the AIEgen properties of TPEPY moiety is quenched by quinone group of MMC via photo-induced electron transfer (PET). TPEPY-S-MMC can be activated by intracellular thiols (GSH/Cys). The higher levels of GSH present in cancer cells could effectively activate TPEPY-S-MMC resulting in the formation of cytotoxic chemotherapeutic drug MMC and TPEPY-SH with restored red fluorescence emission and produce $^1O_2$ upon photo-irradiation (Fig. 8A). To study the role of intracellular GSH on the activation of the probe, cells pretreated with GSH inhibitor, BSO is treated with TPEPY-S-MMC that exhibits weak fluorescence emission and negligible toxicity. N.D.P. Singh and co-workers developed a photo-responsive fluorescent nanoparticles $TPE(Cbl)_4NPs$.[29] The design strategy comprised of covalently conjugated four equivalents of chemotherapeutic drug chlorambucil (Cbl) with AIEgen TPE. The fluorescence emission of $TPE(Cbl)_4NPs$ is owing to its AIE characteristic, the activation of the $TPE(Cbl)_4NPs$ sequentially releases Cbl and photoproduct $TPE(OH)_4NPs$ (Fig. 8B). In vitro studies demonstrated the AIE behavior is important for its photoactivity, on photolysis the fluorescence emission starts decreasing due to the release of Cbl and $TPE(OH)_4NPs$, which can act as self-reporter for drug activation. MTT assay results showed that $TPE(Cbl)_4NPs$ exhibit potent photocytotoxicity against HeLa cells when compared with four equivalence of Cbl and $TPE(OH)_4NPs$.

G. Feng *et al.* reported a light-up fluorescence probe TPETH-Mito-1ART (Fig. 8C), that can be able to deliver both artemisinin (ART) and AIE-PS.[30] The probe is made up of TPETH as AIE core, two arms of the quaternary ammonium salts to target cancer cells mitochondria, one or both the arms of quaternary ammonium methyl group is replaced with ART. The bioactivities of ART have been reported recently; heme is

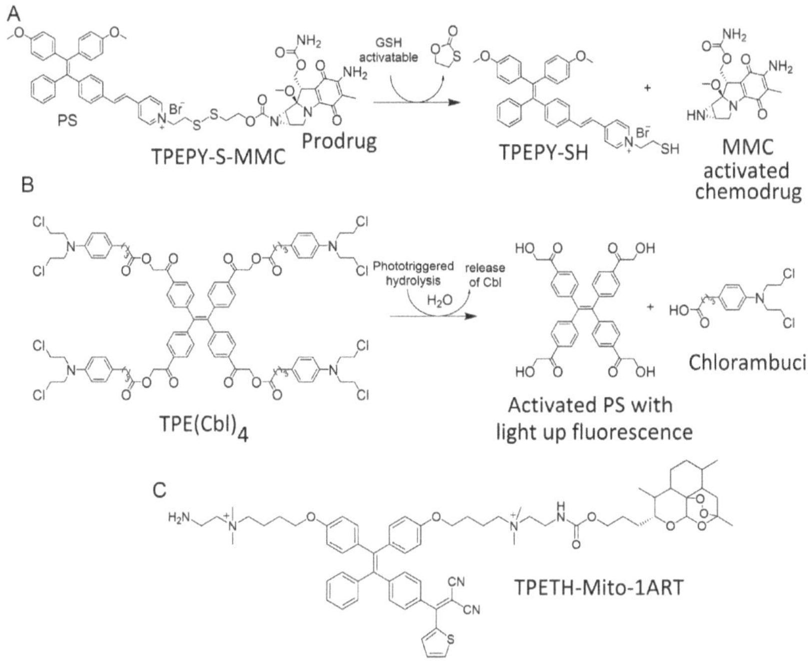

Fig. 8 AIE-PSs based prodrugs. (A) GSH activatable TPEPY-S-MMC for the delivery of chemotherapeutic agent Mitomycin C. (B) Photo-triggered delivery of chlorambucil from TPE(Cbl)$_4$. (C) Mitochondrial heme activated delivery of artemisinin from TPETH-Mito-1ART.

known to activate ART, which could further inhibit the functioning of anti-apoptotic factors. As heme is produced in mitochondria, they have developed mitochondria targeting the AIE-PS probe with ART. The probe, TPETH-Mito-1ART shows the least toxicity under dark condition, upon photo-irradiation with white light it exhibit potent phototoxicity, whereas when they have studied the anticancer activity of ART or TPETH-Mito separately they showed the least phototoxicity, indicating the important role of this combination strategy in synergistically improving the anticancer activity against cancer cells. Simultaneously, when both the arms are functionalized with ART, TPETH-Mito-2ART, it shows potent cytotoxicity even under dark condition, which was evident from its photocytotoxic index (PI). The poor selectivity of TPETH-Mito-2ART may be harmful to normal cells and thus curtail its practical utility. Interestingly, this combination strategy has drastically reduced cancer cell migration, which mitigates cancer spreading to other organs.

7. Polymer based AIE-PS for PDT

Mitochondria-targeted AIE polymer (Fig. 9A) was synthesized, tethered with upconversion nanoparticles (UCNPs) to produce UCNPs@PAIE-TPP NPs.[31] The pH-sensitive polyethylene glycol (PEG) is introduced on the surface of the NPs to yield UCNPs@PAIE-TPP-PEG NPs. The role of PEG group is to reduce the non-specific interactions between proteins and NPs. The selective cleavage of PEG group was envisaged at acidic tumor environment and releases positively charged NPs that can target cancer cell mitochondria. The rationale for designing AIE polymer functionalized UCNPs is to activate the AIE-PS (PAIE) with less harmful and highly penetrating NIR radiation. Irradiation of UCNPs@PAIE-TPP-PEG NPs with 980 nm laser produces ROS, which was ascertained from DCF-DA assay. The NPs show good biocompatibility under dark condition, upon NIR

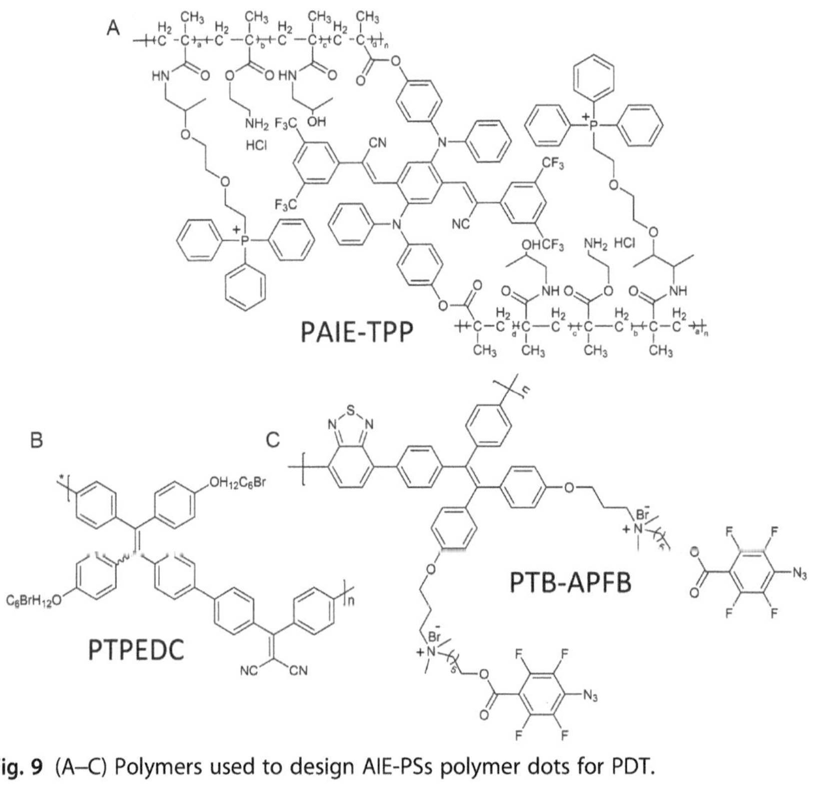

Fig. 9 (A–C) Polymers used to design AIE-PSs polymer dots for PDT.

irradiation, it exhibits potent cytotoxicity against A549 cells that they confirmed from MTT assay. B. Liu and co-workers synthesized two AIE-PS polymers, PTPEDC1 and PTPEDC2 (Fig. 9B) with efficient $^1O_2$ production capability and high 2PA cross-section for TP-PDT.[32] TD-DFT calculation shows the reduction in energy gap between S_1 and T_1 excited states (ΔE_{S1-T1}) and the energy levels are becoming denser after polymerization, indicating the effective ISC, facilitate $^1O_2$ generation. The energy levels of PTPEDC2 are much denser than PTPEDC1 reveal that PTPEDC2 can act as an efficient photosensitizer. The AIE-PS polymer dots were further functionalized with cell penetrating peptide (TAT-SH) to enhance their internalization. Treatment of AIE-TAT PS polymer dots with HeLa cells show negligible dark toxicity, upon irradiation with 820 nm femtosecond laser for different scans increases the cytotoxicity. Fluorescein diacetate and propidium iodide staining of the irradiated cells were used to differentiate live/dead cells using CLSM. The red dots were observed at the laser focal point, indicating the high precision of TP-PDT. PTPEDC2-TAT exhibit potent phototoxicity in vitro and in vivo zebrafish model when compared with PTPEDC-TAT and PTPEDC1-TAT, due to its high 2PA cross-section and better $^1O_2$ generation efficiency.

AIE-PS polymer PTB-APFB was designed and synthesized to target pathogenic bacteria over mammalian cells selectively.[33] PTB-APFB polymer is made up of AIE active tetraphenylethylene (TPE) and benzothiadiazole (Fig. 9C). The integral parts of polymer contain D-π-A skeleton, quaternary ammonium ion and 4-azidoperfluorobenzoate (APFB). PTB-APFB bind with the cell walls of different pathogenic microorganisms such as *Candida albicans*, *S. aureus* and *E. coli*. In contrast its internalization into the cytoplasm was observed only in the dead microorganisms. The difference in the localization of PTB-APFB could be useful to discriminate live/dead microorganisms. Most notably, PTB-APFB exhibits weak binding affinity towards mammalian cells. Moreover, when HeLa cells and microorganisms were cultured together and further treatment with PTB-APFB, selective internalization was realized from the fluorescence emission on the microorganisms, whereas no noticeable fluorescence emission was observed from HeLa cells. The effective growth inhibition of pathogenic microorganisms after treatment with PTB-APFB under light irradiation was tested in vitro and in vivo. The quicker recovery from the infection was observed with PTB-APFB when compared with the known anti-microbial agents, cefalotin.

8. Nanoparticles (NPs) based AIE-PS for PDT

The first report of organic nanoparticles with targeted PDT activity and AIE characteristics was reported by Prof. B. Liu's group.[34] The synthesis of AIE-PS nanoparticles was carried out by using nano-precipitation technique (Fig. 10A), where they have designed a new AIEgen TTD (Fig. 10B), encapsulated it with 1,2-distearoyl-sn-glycero-3-phosphoethanolamine-*N*-[maleimide(poly(ethyleneglycol))-2000] (DSPE-PEG-MAL). Then the surface of the NPs was further modified with cRGD peptide to produce T-TTD NPs. It selectively targets cancer cells that overexpress $\alpha_v\beta_3$ integrin. The CLSM studies have shown that T-TTD NPs are high localized in MDA-MB-231 cells that overexpress $\alpha_v\beta_3$ integrin and thus exhibit excellent PDT activity. In contrast, very weak fluorescence signals were observed on MCF-7 and NIH 3T3 cells due to their low-level expression of $\alpha_v\beta_3$ integrin resulting in the least phototoxicity.

A similar strategy was used to design one step formulation of T-TTD nanodots, studied its targeted theranostic activity against cholangiocarcinoma and hepatocellular carcinoma (HCC), respectively.[35] Dual targeted organic nanodots were synthesized for image-guided PDT applications.[36] A new AIEgen, DPBA-TPE (Fig. 10C) was synthesized and encapsulated inside the nanoparticles using a modified nanoprecipitation procedure. The nanoparticles surface was appended with folic acid (FA), triphenylphosphine (TPP) to produce FA-AIE-TPP dots, which targets cancer cells that over express folate receptor (FR) and subsequent mitochondrial localization. FA-AIE-TPP dots produces ROS upon white light irradiation, depolarizes the mitochondrial membrane, reduces cancer cell migration and leads to cell apoptosis.

In later studies, they have designed and synthesized efficient AIE-PS, TPETCAQ (Fig. 10D) with excellent photostability and NIR emission characteristics.[37] The nanoparticles were prepared using the nano-precipitation method as mentioned above, further the surfaces of NPs were modified with RKKRRQRRRC peptide derived from transactivator of transcription (TAT) protein encoded by human immunodeficiency virus (HIV-1) to yield TPETCAQ NPs with improved cell penetration ability, which was confirmed from cellular uptake studies. The protective polymer matrix reduces the exposure of TPETCAQ to polar media, thus resulted in bright fluorescence and high $^1O_2$ generation efficiency of TPETCAQ NPs. In vitro and in vivo experiments clearly signify the potent anticancer activity of

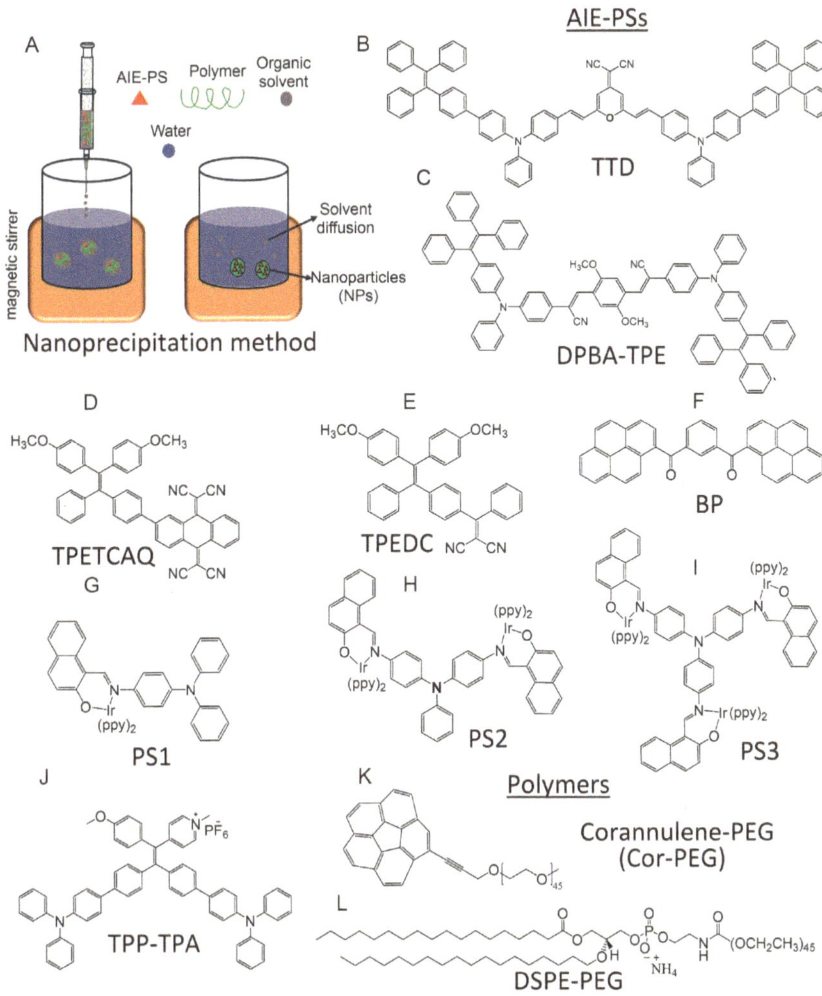

Fig. 10 (A) Nanoprecipitation method to synthesize AIE-PS encapsulated nanoparticles (NPs). (B–J) Chemical structure of AIE-PSs used to design NPs. (K, L) Polymers used to design NPs.

TPETCAQ NPs under light irradiation, whereas it exhibits negligible dark toxicity. In vivo mouse model studies shown the reduction in the tumor size after treatment with TPETCAQ NPs under light irradiation, but no noticeable weight losses were observed even after 10 days, implying its safety profile for ensuing clinical applications. A similar strategy was utilized to synthesize TPA-AIE dots, a new AIEgen, TPEDC (Fig. 10E) with a high TPA cross-section and enhanced ROS generation efficiency was chosen.[38] Then it was encapsulated with DSPE-PEG-MAL and functionalized with HIV-1 TAT peptide to produce T-TPEDC dots. To study the TP-PDT activity of the T-TPEDC dots, they were treated with HeLa cells followed by photo-irradiation were performed using 800 nm laser. Increasing the number of scans increases propidium iodide (PI) positive staining and fluorescein isothiocyanate (FITC)-labeled Annexin V staining in the scanned region, that signifies the necrotic and apoptotic cell death, whereas the cells outside the scanned areas are unaffected, indicating its precise targeting ability and excellent biocompatibility.

Recently, bis(pyrene) based AIE-PS was developed (Fig. 10F) and entrapped inside liposomes to produce BP@liposomes.[39] The photosensitivity of BP is lost after liposomal entrapment, but the photosensitivity can be regained back once the liposomes are decomposed at the targeted site by phospholipases. The PDT activity of BP and BP@liposomes were studied, BP shows potent toxicity in the presence of light (OPA/TPA), whereas BP@liposomes are not toxic even after light irradiation. The decomposition of liposomes takes place slowly at the target site, after 30 h of treatment, the release of BP and reaggregation was noted, subsequent light irradiation shows excellent PDT activity. This strategy may play an imperative role in controlling the activity of AIE-PS and liberate the darkroom treatment for PDT.

A series of red light-emitting mononuclear, binuclear and trinuclear cyclometallated iridium complexes with AIE characteristics were developed.[40] The rationale behind the attempt to synthesize multinuclear iridium complex is to improve the metal-to-ligand charge transfer (^{3}MLCT), increase the molar extinction coefficient and improve $^1O_2$ generation efficiency. The iridium complexes (PS1, PS2, and PS3) (Fig. 10G–I) were encapsulated inside DSPE-PEG-MAL, further functionalized the NPs surfaces with HIV-1 TAT peptide to produce of PS1-PS3 NPs. The photoluminescence quantum yield and the excited state lifetime of PS1-PS3 NPs are much higher than their corresponding Ir(III) complexes. Therefore, the NPs can act as a better photosensitizer and light up fluorescence probe for image-guided PDT applications. PS3-NPs show potent phototoxicity and appears as a promising lead molecule for PDT, owing to its longer excited state lifetime, high photoluminescence quantum yield, efficient $^1O_2$ generation ability and least dark toxicity.

The effect of encapsulating matrix on the efficacy of AIE-PS dots has been studied by B. Z. Tang and co-workers.[41] A new AIE-PS, TPP-TPA (Fig. 10J) has been designed and the encapsulated using Cor-PEG

(Fig. 10K) and DSPE-PEG (Fig. 10L). The modification of the encapsulating matrix from DSPE-PEG to Cor-PEG resulted in enhanced fluorescence emission and improved ROS generation ability. Cor-PEG-AIE dots improved activity is due to increased intra-particle rigid environment, strong interaction between corannulene core and TPP-TPA, thereby suppressing the non-radiative decay and facilitating ISC. This strategy improves AIE dots efficiency for NIR image-guided PDT and provides a platform to develop superior fluorescent NPs for image-guided surgery and other biomedical applications.

9. AIE-PS nanoparticles for combination PDT

The combination of chemotherapy with PDT is an emerging strategy for cancer treatment. X. Yi *et al.* developed a reduction-sensitive polymeric NPs, TB@PMP, for combination therapy.[42] TB@PMP was prepared by reacting reduction-sensitive chemotherapeutic agent, paclitaxel functionalized polymeric prodrug (PMP) (Fig. 11A) with red emissive AIEgen, TB using dialysis method. TB@PMP micelles exhibit bright fluorescence emission and high photosensitization ability upon light irradiation owing to encapsulated TB. The release of paclitaxel from TB@PMP takes place under intracellular reducing conditions commonly observed in cancer cells, facilitating the disassembly of microtubule and induced cell apoptosis. The biological studies carried on HeLa cells and tumor-bearing mouse model show the drastic reduction in tumor size after treatment with TB@PMP followed by light irradiation. The synergistic combination effect of PDT and chemotherapeutic activities of TB@PMP micelles upon irradiation exhibit potent anticancer activity against HeLa cells. In another strategy, they have developed a DEB/TQR@PMP micelles by integrating DEB (Fig. 11B), drug resistance inhibitor tariquidar (TQR) (Fig. 11C) and reduction-sensitive paclitaxel functionalized polymeric prodrug (PMP).[43] The synergistic enhancement in the anticancer activity was noted when DEB/TQR@PMP micelles were treated with HeLa cells, SKOV-3 cells and drug-resistant SKOV-3 cells followed by light irradiation. The better efficacy of these micelles against drug resistant SKOV-3 cells is due to the release of TQR, which inhibits the overexpressed P-glycoprotein from drug resistant SKOV-3 cells and sensitive to the treatment. In multidrug resistance, SKOV-3 cells bearing tumor mouse model, DEB/TQR@PMP micelles reverse the MDR effect and inhibit the tumor growth upon infrared

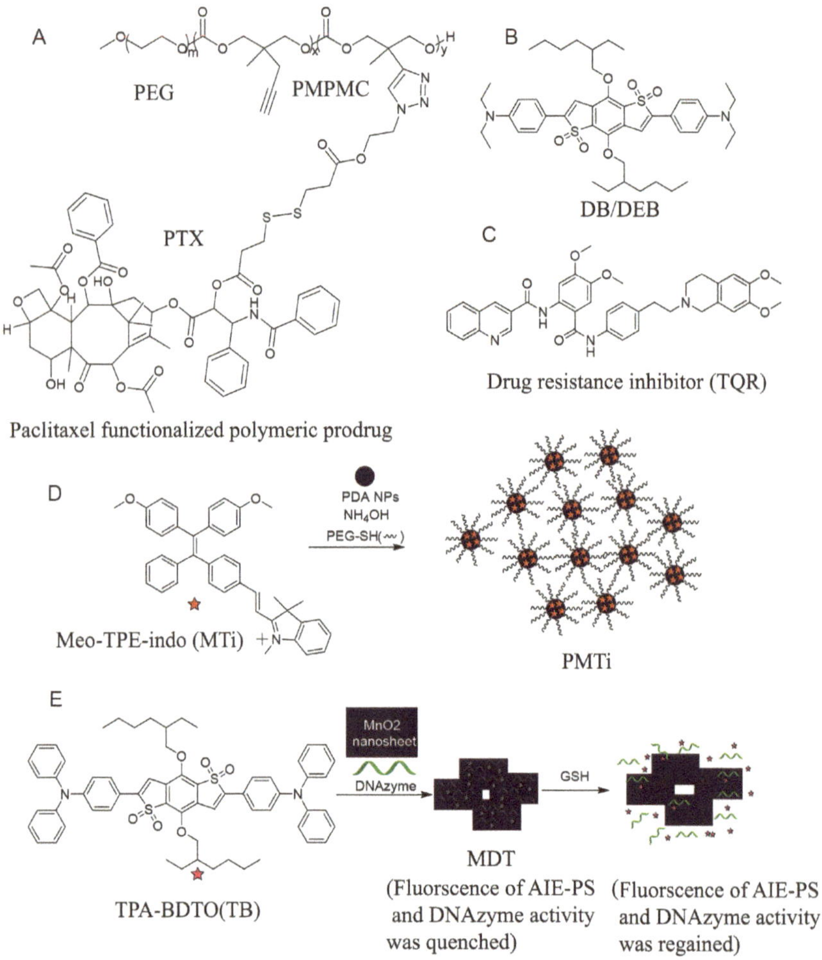

Fig. 11 (A–C) Chemical structures of polymeric prodrug containing paclitaxel (PEG-*b*-PMPMC-*g*-PTX), drug resistance inhibitor (TQR) and AIE-PS (DEB) used to design DEB/TQR@PMP micelles for combination therapy (PDT and chemotherapy). (D) AIE-PS (MTi) encapsulated PEG-PDA NPs for PDT and PTT. (E) AIE-PS (TB) and DNAzyme modified MnO_2 nanosheets for targeted gene delivery and PDT.

irradiation. The high TPA and good photostability of these micelles make it a potential image-guided combination therapy to overcome multidrug resistance.

Y. Chen *et al.* designed and synthesized a new AIE-PS, MTi (MeO-TPE-indo), which has an electron-donating methoxy group and an electron-withdrawing indolium group on TPE.[44] The presence of a cationic indolium group on MTi facilitates the accumulation of MTi selectively on mitochondria. To improve the dispersion of MTi in water and enhance the retention at the tumor site, MTi has been attached to the surface of polydopamine nanoparticles (PDA NPs). The PMTi (PDA-MeO-TPEindo) nanocomposite was prepared by attaching the surface of PDA NPs with MTi, and this assembly is enabled by electrostatic, π-π, and hydrogen bonding interactions (Fig. 11D). Also, the photothermal conversion efficacy of PDA can be exploited for photothermal therapy (PTT). The synthesized PMTi nanocomposite triggers apoptosis in cancer cells, inhibiting tumor growth in mouse models upon white light and NIR irradiation. The fluorescence characteristics of PMTi nanocomposite are maintained for a longer time, make it as multifunctional diagnosis and treatment agent for image-guided PDT and PTT.

Multifunctional AIE-PS organic nanoparticles functionalized with siRNA has been developed for targeted PDT and RNA interference therapy.[45] The newly synthesized AIE-PS, TTD was encapsulated with reduction/GSH sensitive siRNA targeting vascular endothelial growth factor (siVEGEF) to yield DSPE-PEG-siVEGEF and $\alpha_v\beta_3$ integrin targeting DSPE-PEG-cRGD matrix to produce cRGD-siVEGF-TTD NPs. The nanoparticles show efficient $^1O_2$ producing ability with a high $^1O_2$ quantum yield upon irradiation. cRGD present on the surface of NPs target overexpressed $\alpha_v\beta_3$ integrin and the role of siVEGEF in the matrix is to interfere with upregulated VEGEF in cancer cells, suppress angiogenesis. The release of siVEGEF is tested in the presence and absence of GSH, the release is favored in the presence of GSH, whereas in the absence of GSH, the release is minimal. The targeting ability of cRGD-siVEGF-TTD NPs was tested against $\alpha_v\beta_3$ integrin overexpressing cancer cells (MDA-MB-231) and low-level $\alpha_v\beta_3$ integrin expressing cancer cells (MCF-7 and SK-BR-3). All the cell lines were treated with cRGD-siVEGF-TTD NPs, the CLSM images show strong fluorescence signals from MDA-MB-231 cells. In contrast, a weak fluorescence signals were observed from MCF-7 and SK-BR-3 cells, indicating the selective targeting ability of the NPs. Cytotoxicity assay was carried to determine the cell viability of MDA-MB-231 cancer cells after treatment with cRGD-siVEGF-TTD NPs, the reduction in the cell viability was observed due to the release of siVEGF and the drastic reduction in cell viability was noted after light irradiation as a result of PDT. The synergistic action of PDT and RNA interference therapy has improved the anticancer activity and open a new avenue for designing multifunctional theranostic agents.

Similarly, a synergistic delivery strategy was developed by X. Wang *et al.*,[46] where a GSH responsive MnO_2 nanosheet was loaded with an AIEgen, TB and DNAzyme through π-π stacking and electrostatic interactions that leads to the formation of MnO_2-DNAzyme-TB nanocomposite (MDT) (Fig. 11E). The fluorescence of TB and the activity of DNAzyme is completely quenched after loading them into MnO_2 nanosheets. The reductive atmosphere in cancer cells reduces the MnO_2 nanosheet and further resulted in the formation of Mn^{2+} ions, facilitating the release of TB and DNAzyme. The released DNAzyme degrade the EGR-1 mRNA in cancer cells and the TB form aggregates, regain its AIE behavior, upon light irradiation it produces $^1O_2$. The expression levels of EGR-1 mRNA were monitored on MCF-7 cell line in control and after treatment with MDT, the low-level expression of EGR-1 mRNA was observed in MDT treated cells signifies the degradation of EGR-1 mRNA by DNAzyme. The good therapeutic response of MDT upon irradiation might be due to the synergistic effect of PDT and gene therapy.

10. Concluding remarks and future directions

Photodynamic therapy has attracted considerable attention recently owing to its non-invasive and controllable treatment modality. The unique characteristics of AIE-PSs improve the ROS generation ability and fluorescence emission properties in the aggregated state, make them as promising molecules for image-guided PDT. In this chapter, we discussed about various design strategies to achieve targeted PDT, 2PA-PDT, polymer/nanoparticle encapsulated PDT and combination PDT. These strategies can improve the biocompatibility and targeting ability of AIE-PSs. Finally, we emphasized the importance of multifunctional AIE-PS NPs for combination PDT. A combination of AIE-PSs with chemotherapy/phototherapy/gene therapy synergistically improves the anticancer activity and combat against multidrug resistance. There are still many challenges that need to be addressed to utilize the full potential of AIE-PSs for PDT applications. Few of them are, in-depth tumor treatment is highly challenging with PDT, designing AIE-PSs with high NIR absorbance and emission, developing micro-endoscopy to image/irradiate a selective portion of organs, scalable and simple strategies to functionalize AIE-PSs with tumor-targeting vectors to target cancer cells over normal cells selectively.

Acknowledgments

V.V. is thankful to DST, New Delhi, India, for financial support under the DST-INSPIRE faculty award scheme [DST/INSPIRE/04/2016/001173] and IIT Roorkee for Faculty Initiation Grant (FIG). V.S. acknowledges a research fellowship from UGC, New Delhi.

References

1. Dougherty TJ, Gomer CJ, Henderson BW, et al. Photodynamic therapy. *J Natl Cancer Inst.* 1998;90:889–905.
2. Celli JP, Spring BQ, Rizvi I, et al. Imaging and photodynamic therapy: mechanisms, monitoring, and optimization. *Chem Rev.* 2010;110:2795–2838.
3. Baskaran R, Lee J, Yang SG. Clinical development of photodynamic agents and therapeutic applications. *Biomater Res.* 2018;22:25.
4. Luo J, Xie Z, Lam JW, et al. Aggregation-induced emission of 1-methyl-1,2,3,4,5-pentaphenylsilole. *Chem Commun.* 2001;18:1740–1741.
5. Yang Z, Yin W, Zhang S, et al. Synthesis of AIE-active materials with their applications for antibacterial activity, specific imaging of mitochondrion and image-guided photodynamic therapy. *ACS Appl Bio Mater.* 2020;3:1187–1196.
6. Hu F, Xu S, Liu B. Photosensitizers with aggregation-induced emission: materials and biomedical applications. *Adv Mater.* 2018;30:1801350.
7. Cai X, Liu B. Aggregation-induced emission: recent advances in materials and biomedical applications. *Angew Chem Int Ed.* 2020;59:9868–9886.
8. Wang D, Lee MMS, Shan G, et al. Highly efficient photosensitizers with far-red/near-infrared aggregation-induced emission for in vitro and in vivo cancer theranostics. *Adv Mater.* 2018;30:1802105.
9. Zhang T, Li Y, Zheng Z, et al. In situ monitoring apoptosis process by a self-reporting photosensitizer. *J Am Chem Soc.* 2019;141:5612–5616.
10. Zou J, Lu H, Zhao X, et al. A multi-functional fluorescent probe with aggregation-induced emission characteristics: mitochondrial imaging, photodynamic therapy and visualizing therapeutic process in zebra fish model. *Dyes Pigments.* 2018;151:45–53.
11. Yang C, Hu R, Lu F, et al. Traceable cancer cell photoablation with a new mitochondria-responsive and -activatable red-emissive photosensitizer. *Chem Commun.* 2019;55:3801–3804.
12. You X, Ma H, Wang Y, et al. Pyridinium-substituted tetraphenylethylene entailing alkyne moiety: enhancement of photosensitizing efficiency and antimicrobial activity. *Chem Asian J.* 2017;12:1013–1019.
13. Dai Y, He F, Ji H, Zhao X, Misal S, Qi Z. Dual-functional NIR AIEgens for high-fidelity imaging of lysosomes in cells and photodynamic therapy. *ACS Sens.* 2020;5:225–233.
14. Zhuang J, Yang H, Li Y, Wang B, Li N, Zhao N. Efficient photosensitizers with aggregation induced emission characteristics for lysosome and Gram-positive bacteria-targeted photodynamic therapy. *Chem Commun.* 2020;56:2630–2633.
15. Zhang W, Huang Y, Chen Y, et al. Amphiphilic tetraphenylethene-based pyridinium salt for selective cell-membrane imaging and room-light-induced special reactive oxygen species generation. *ACS Appl Mater Interfaces.* 2019;11:10567–10577.
16. Xu W, Lee MMS, Lee MMS, et al. Facile synthesis of AIEgens with wide color tunability for cellular imaging and therapy. *Chem Sci.* 2019;10:3494–3501.
17. Xu W, Lee MMS, Nie JJ, et al. Three-pronged attack by homologous far-red/NIR AIEgens to achieve 1 + 1 + 1 > 3 synergistic enhanced photodynamic therapy. *Angew Chem Int Ed.* 2020;59:9610–9616.

18. Jiang M, Kwok RTK, Li X, et al. A simple mitochondrial targeting AIEgen for image-guided two-photon excited photodynamic therapy. *J Mater Chem B.* 2018;6: 2557–2565.
19. Zheng Z, Liu H, Zhai S, et al. Highly efficient singlet oxygen generation, two photon photodynamic therapy and melanoma ablation by rationally designed mitochondria specific near-infrared AIEgens. *Chem Sci.* 2020;11:2494–2503.
20. Kong Q, Ma B, Yu T, et al. Two-photon AIE fluorophore as a photosensitizer for highly efficient mitochondria-targeted photodynamic therapy. *New J Chem.* 2020;44: 9355–9364.
21. Liu J, Jin C, Yuan B, et al. Selectively lighting up two-photon photodynamic activity in mitochondria with AIE-active iridium(III) complexes. *Chem Commun.* 2017;53: 2052–2055.
22. Qiu K, Ouyang M, Liu Y, et al. Two-photon photodynamic ablation of tumor cells by mitochondria-targeted iridium(III) complexes in aggregate states. *J Mater Chem B.* 2017;5:5488–5498.
23. He Z, Gao Y, Zhang H, et al. Platinum-AIEgen coordination complex for imaging-guided annihilation of cisplatin-resistant cancer cells. *Chem Commun.* 2020;56: 7785–7788.
24. Hu F, Huang Y, Zhang G, Zhao R, Yang H, Zhang D. Targeted bioimaging and photodynamic therapy of cancer cells with an activatable red fluorescent bioprobe. *Anal Chem.* 2014;86:7987–7995.
25. Yuan Y, Xu S, Zhang CJ, Zhang R, Liu B. Dual-targeted activatable photosensitizers with aggregation-induced emission (AIE) characteristics for image-guided photodynamic cancer cell ablation. *J Mater Chem B.* 2016;4:169–176.
26. Yuan Y, Zhang CJ, Kwok RTK, et al. Light-up probe for targeted and activatable photodynamic therapy with real-time in situ reporting of sensitizer activation and therapeutic responses. *Adv Funct Mater.* 2015;25:6586–6595.
27. Yuan Y, Zhang C, Gao M, Zhang R, Tang BZ, Liu B. Specific light-up bioprobe with aggregation-induced emission and activatable photoactivity for the targeted and image-guided photodynamic ablation of cancer cells. *Angew Chem Int Ed.* 2015;54: 1780–1786.
28. Hu F, Yuan Y, Mao D, Wu W, Liu B. Biomaterials smart activatable and traceable dual-prodrug for image-guided combination photodynamic and chemo-therapy. *Biomaterials.* 2017;144:53–59.
29. Parthiban C, Pavithra M, Reddy LVK, Sen D, Singh NDP. Single-component fluorescent organic nanoparticles with four-armed phototriggers for chemo-photodynamic therapy and cellular imaging. *ACS Appl Nano Mater.* 2019;2:3728–3734.
30. Feng G, Liu J, Zhang C, Liu B. Artemisinin and AIEgen conjugate for mitochondria-targeted and image-guided chemo and photodynamic cancer cell ablation. *ACS Appl Mater Interfaces.* 2018;10:11546–11553.
31. Guan Y, Lu H, Li W, et al. Near-infrared triggered upconversion polymeric nanoparticles based on aggregation-induced emission and mitochondria targeting for photodynamic cancer therapy. *ACS Appl Mater Interfaces.* 2017;9:26731–26739.
32. Wang S, Wu W, Manghnani P, et al. Polymerization-enhanced two-photon photosensitization for precise photodynamic therapy. *ACS Nano.* 2019;13:3095–3105.
33. Zhou T, Hu R, Wang L, et al. An AIE-active conjugated polymer with high ROS-generation ability and biocompatibility for efficient photodynamic therapy of bacterial infections. *Angew Chem Int Ed.* 2020;59:9952–9956.
34. Yuan Y, Feng G, Qin W, Tang BZ, Liu B. Targeted and image-guided photodynamic cancer therapy based on organic nanoparticles with aggregation-induced emission characteristics. *Chem Commun.* 2014;50:8757.

35. Li M, Gao Y, Yuan Y, et al. One-step formulation of targeted aggregation-induced emission dots for image-guided photodynamic therapy of cholangiocarcinoma. *ACS Nano*. 2017;11:3922–3932.
36. Feng G, Qin W, Hu Q, Tang BZ, Liu B. Cellular and mitochondrial dual-targeted organic dots with aggregation-induced emission characteristics for image-guided photodynamic therapy. *Adv Healthc Mater*. 2015;4:2667–2676.
37. Wu W, Mao D, Hu F, et al. A highly efficient and photostable photosensitizer with near-infrared aggregation-induced emission for image-guided photodynamic anticancer therapy. *Adv Mater*. 2017;29:1700548.
38. Gu B, Wu W, Xu G, et al. Precise two-photon photodynamic therapy using an efficient photosensitizer with aggregation-induced emission characteristics. *Adv Mater*. 2017;29: 1701076.
39. Yang Y, Wang L, Cao H, et al. Photodynamic therapy with liposomes encapsulating photosensitizers with aggregation-induced emission. *Nano Lett*. 2019;19:1821–1826.
40. Zhang L, Li Y, Che W, et al. AIE multinuclear Ir(III) complexes for biocompatible organic nanoparticles with highly enhanced photodynamic performance. *Adv Sci*. 2019;6:1802050.
41. Gu X, Zhang X, Ma H, et al. Corannulene-incorporated AIE nanodots with highly suppressed nonradiative decay for boosted cancer phototheranostics in vivo. *Adv Mater*. 2018;30:1801065.
42. Yi X, Dai J, Han Y, et al. A high therapeutic efficacy of polymeric prodrug nano-assembly for a combination of photodynamic therapy and chemotherapy. *Commun Biol*. 2018;1:202.
43. Zhen S, Yi X, Zhao Z, Lou X, Xia F, Tang BZ. Biomaterials drug delivery micelles with efficient near-infrared photosensitizer for combined image-guided photodynamic therapy and chemotherapy of drug-resistant cancer. *Biomaterials*. 2019;218:119330.
44. Chen Y, Ai W, Guo X, et al. Mitochondria-targeted polydopamine nanocomposite with AIE photosensitizer for image-guided photodynamic and photothermal tumor ablation. *Small*. 2019;15:1902352.
45. Jin G, Feng G, Qin W, Tang BZ, Liu B, Li K. Multifunctional organic nanoparticles with aggregation-induced emission (AIE) characteristics for targeted photodynamic therapy and RNA interference therapy. *Chem Commun*. 2016;52:2752–2755.
46. Wang X, Dai J, Wang X, et al. MnO_2-DNAzyme-photosensitizer nanocomposite with AIE characteristic for cell imaging and photodynamic-gene therapy. *Talanta*. 2019;202:591–599.

CHAPTER FOUR

Aggregation induced bright organic luminogens: Design strategies, advanced bio-imaging and theranostic applications

Debasish Barman[a], Retwik Parui[a], Kavita Narang[a], Rajdikshit Gogoi[a], and Parameswar Krishnan Iyer[a,b,c,*]

[a]Department of Chemistry, Indian Institute of Technology Guwahati, Guwahati, Assam, India
[b]Centre for Nanotechnology, Indian Institute of Technology Guwahati, Guwahati, Assam, India
[c]School of Health Sciences and Technology, Indian Institute of Technology Guwahati, Guwahati, Assam, India
*Corresponding author: e-mail address: pki@iitg.ac.in

Contents

Progress in Molecular Biology and Translational Science, Volume 185
ISSN 1877-1173
https://doi.org/10.1016/bs.pmbts.2021.07.001

Abstract

The concept of aggregation-induced emission (AIE) in purely organic luminescent molecules has drawn wide attention in the last two decades. Despite the many challenges, AIE-probes have opened versatile opportunities in many research fields. In particular, the emerging functional properties of room temperature phosphorescence (RTP) and thermally activated delayed fluorescence (TADF) have boosted the unique features of AIE luminogens (AIEgens). Thus, these luminescent materials extended the utility in sensing, imaging, optoelectronics and theranostic applications in biological field over the conventional fluorescent probe. Unlike the sensitivity of triplet state by oxygen and moisture, the long-lived phosphorescence and delayed fluorescence resulting from the enhanced intersystem crossing (ISC) and reverse intersystem crossing (RISC) from excited triplet state (T_1) to excited singlet state (S_1) in these luminophores gives rise to long lifetimes ranging from nanoseconds to milliseconds even up to seconds. As compared to traditional fluorescence molecules advanced AIE probes show high contrast imaging and deeper penetration depth, which have been demonstrated through near infrared I and II (NIR-I & NIR-II) fluorescence imaging, room temperature after-glow imaging and photoacoustic imaging. This chapter highlights the recent developments and principle of the efficient design of AIE probe with multi-functional properties evolved with new strategies for translational applications via fluorescence imaging, photoacoustic imaging and image-guided photodynamic/photothermal therapy (PDT/PTT) including future opportunities for AIEgens to advance the overall biomedical field.

Graphical abstract

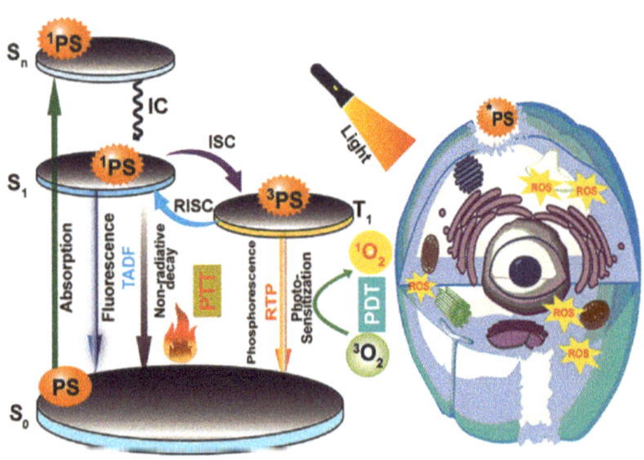

1. Introduction

Fluorescence imaging technique has flourished as an indispensable tool in the area of biomedicine for the visualization of specific bioanalytes with high-resolution, observation of complicated biological processes, and theranostic applications.[1–3] The high sensitivity, biocompatibility, safety, low cost, and non-invasiveness are encouraging evidences on the multimodal capability of new generation materials for modern healthcare based applications and have fascinated growing interest in clinical and pre-clinical studies (both in vivo and in vitro).[4] In the last decade, efficient emitters such as organic dyes,[5] lanthanide metal-based nanoparticles,[6,7] conjugated polymers[8] inorganic quantum dots[9,10] etc. have been developed. Organic dyes and nanoparticles doped with organic dyes are superior alternatives for imaging due to their impressive biocompatibility, condensed state properties and well-suited synthetic processes.[4] Organic small molecule-based emitters possess high fluorescence quantum yields, excellent biocompatibility, good photostability, and lesser phototoxicity during biological application.[11,12] The hydrophobic nature of most of the organic dyes helps them to form aggregates naturally in a hydrophilic biological system and permits aggregation caused quenching (ACQ) of the fluorophores due to π-π stacking.[13] This drawback has been overcome after the concept of aggregation induced emission (AIE) was introduced by Ben Zhong Tang group in 2001 that provided a new direction to the field of fluorescence imaging using AIEgens as candidates for external contrast (exogenous) agents.[14]

Inside a biological system, AIEgens may bind with different biomolecules present in the system and due to the influence of the surroundings, the free intramolecular motions of the molecules is constrained, which enhances the radiative decay of the luminophores.[15] By analyzing the nature and intensity of the aggregate emission, some biomolecules and their certain interactions inside the system can be traced.[16] The spontaneous aggregation of hydrophobic dyes inside the hydrophilic bio-environment enables AIEgens to easily "turn-on" followed by impressive fluorescence quantum yield which is beneficial for improving the spatial resolution of imaging with better contrast.[17]

In recent years significant efforts have been devoted for the advancement of efficient design principles and realizing new and exciting photophysical processes such as, development of near-IR AIEgens with room temperature phosphorescence (RTP)[18] and thermally activated delayed fluorescence

(TADF).[19] The collective properties of such family of molecules enables the probe to be more potential for eliminating tissue autofluorescence, improved signal to noise ratio by facilitating a proper delay time between the long-lived emission of dyes and the pulsed excitation light.

1.1 Organic near-IR AIE molecules

Small organic molecules having emission in the visible region (400–700 nm) have experienced severe complications due to the short-wavelength excitation that are very harmful to living organs and tissues.[20] These visible region emitters also have other limitations like less penetration ability inside the cell and high background autofluorescence. To alleviate this problem, researchers have developed NIR I [(700–900 nm) and NIR II (1000–1700 nm)] organic materials, which displayed less background autofluorescence, low photon scattering, spatial resolution for the visualization of biological complexity and also deep tissue penetration ability.[21,22] (Fig. 1A).

1.2 Organic AIE-RTP molecules

Recently, organic RTP molecules have drawn significant attention in biomedical applications owing to their bright and long RTP emission, good photo-stability, long wavelength emission, good bio-compatibility, water

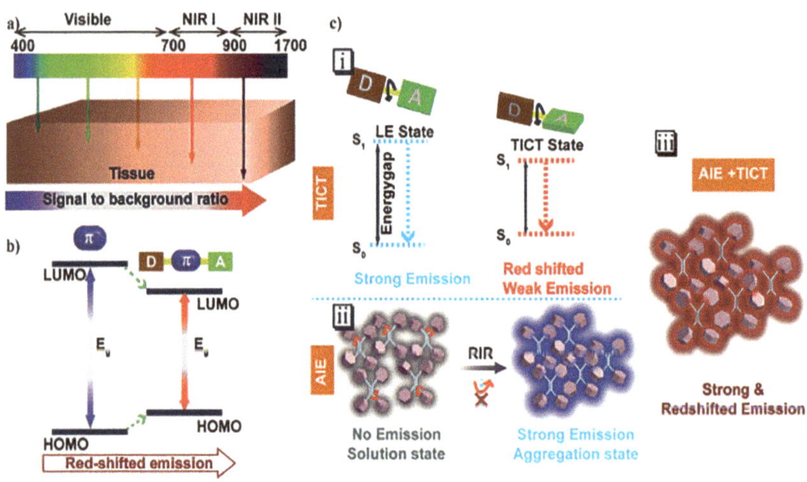

Fig. 1 (A) Graphical representation of the advancement of NIR probe toward biological application. (B) D-A engineering strategy for the reduction of the bandgap. (C) TICT mechanism (i) AIE mechanism, combination of AIE and TICT to achieve red-shifted and highly intense fluorescence emission (ii).

dispensability, suitable size to cross the membrane.[23,24] A few nanoprobes with long-lifetime phosphorescence emission have been reported for optical bioimaging but most of the reported phosphorescent probes comprise of rare earth heavy metals (europium,[25] iridium[26] etc.), which suffer from the potential drawback of toxicity.[27] Hence, development of pure organic RTP materials has gained pace in recent years due to their wide application in organic electronics, sensing, information encryption, security and biological imaging, and tunable luminescence properties.[28] At the same time the photoluminescence (PL) properties of these organic molecules can be fine-tuned by varying the energy levels, electronic configuration and lifetime of the excitons.[29] Typical fluorescence molecules lack certain properties in terms of long term imaging efficiency. Conventional fluorophores can achieve only 25% internal quantum efficiency, whereas remaining 75% excitons are triplet excitons and remain unused in radiation.[30] Thus, there is an urgent necessity to harvest this 75% triplet non-radiative decay to achieve maximum imaging efficacy. Even though, the quenching tendency of the triplet excitons of phosphorescent luminogens by oxygen and water creates difficulty for their use in biological systems effectively, RTP molecules are very attractive in biological imaging due to their longer lifetimes of milliseconds to seconds, and can suppress the auto-fluorescence for several nanoseconds of the tissue.[31–33]

1.3 Organic AIE-TADF molecules

Recently, another class of fluorescence molecules called TADF emitters which display slightly delayed emission due to the involvement of intersystem crossing (ISC) and reverse intersystem crossing (RISC) throughout the process have achieved higher lifetime of up to several microseconds.[34] However, conventional fluorescence imaging suffers from the issue of high tissue autofluorescence, leading to create complications in separating target emission signal from the inherent background autofluorescence.[35] Usually fluorophores with NIR wavelength emissions are preferred as they circumvent the background emission interferences. But NIR organic dyes have the limitation of small stokes shift with large overlap between absorption band and emission band[36,37] that leads them to reabsorption of emitted photon results and unanticipated weak emission. Hence, luminogens with higher emission lifetimes can be a promising alternative for eliminating the comparatively short-lived background fluorescence by time-resolved fluorescence imaging (TRFI).[38] As AIE-TADF (or TADF only) luminogens have higher lifetime (μs) compared to typical fluorescence (ns), they are expected to have the potential capability of eliminating tissue autofluorescence with improved signal to noise ratio by introducing a proper delay time between the long-lived emission of dyes and the pulsed excitation light.[39]

2. Basic aspects and design strategy

2.1 NIR-AIE organic material

Design of organic NIR emitting materials are mostly focused on the reduction of the band gap (E_g) which can be obtained by following two methods: (1) increment of π-conjugation length and (2) engineering of acceptor (A)-donor (D) groups.[40,41] Based on the molecular orbital theory (MOT), the more the π-conjugation, the lesser will be the bandgap between the highest occupied molecular orbital (HOMO) and lowest unoccupied molecular orbital (LUMO) electronic level. Similarly, tuning of the electron donor group can raise the HOMO energy level, and the electron-withdrawing group can lower the LUMO energy level and results in the lesser band gap as well as good NIR emitting probe (Fig. 1B).

Engineering of D-A with π-spacer can diminish the band gap between the HOMO and LUMO and produce NIR emitting molecules. On the other part, flexibility and rigidity of the AIE molecules have a huge aspect in controlling the fluorescence quenching and enhancement respectively. Amalgamation of these two strategies can generate NIR emissive AIE materials endowing excellent photostability and high quantum yield emission in the biological medium. There is another process called twisted intramolecular charge transfer (TICT) that implicates conformation of a single luminophore.[42] This phenomenon is commonly found in the D-A system with a twisted backbone structure. In polar solvents like water, the molecule takes shape as high dihedral angled structure, gaining higher stable and relaxed conformation and the D to A intramolecular CT undergoes very fast in the twisted form. Higher stability of the structure also facilitates emission from the dark TICT state which is much lower than the local excited state (LE) (first excited state for strained coplanar structure). Therefore, TICT molecules display red-shifted emission in a polar solvent. On the other hand, due to several non-radiative decays from TICT state, emission intensity becomes somewhat weakened.

2.1.1 Boosting strategies for high QY NIR II

Apart from extended conjugation, a most common method to achieve more hypochromic emission is strengthening the D and A ability. Improvement in

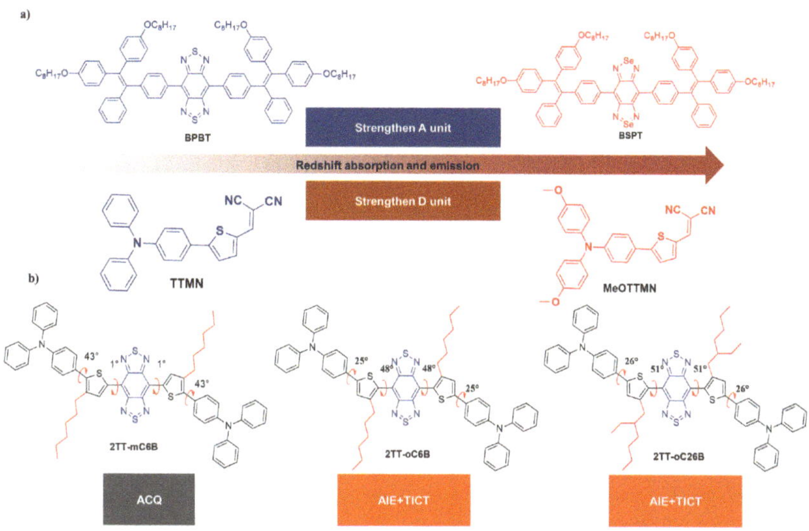

Fig. 2 (A) Molecular design strategy to boost emission toward the NIR II region. Strengthening donor strength and acceptor affinity to tune the bandgap as well as emission. (B) Structural backbone modulation to unveil both AIE and TICT features in the aggregated state.

the donating strength and accepting affinity reduces the energy bandgap. For example, triphenylamine (TPA) unit acts as a good electron donor along with AIE characteristics because of the three rotatable phenyl groups. Presence of TPA unit in TTMN molecule (Fig. 2A) endowed great AIE nature with absorption and emission at 483 nm and 672 nm respectively.[43] However, the introduction of methoxy group in the TPA unit boosted the donating affinity of the TPA group, as a consequence much more red-shifted absorption (499 nm) and emission (701 nm) in the MeOTTMN molecule was observed. From the theoretical study, it was observed that the HOMO-LUMO band gap also decreased from 2.55 to 2.40 eV.

On the other hand, the improvement of acceptor property can also help to modify the bandgap. For example, Benzo[1,2-c:4,5-c']bis([1,2,5] thiadiazole) (BBTD) is a well-known acceptor unit used in many NIR emitting materials for its strong accepting nature. Attaching TPE unit (weak electron donor) to BBTD gave good AIE behavior as well as low bandgap of the probe.[44] BPBT displayed emission maxima at 810 nm that was red-shifted to 897 nm for BSPT molecule. This red shifting of emission wavelength resulted due to increment of the electron deficiency of BBTD group through incorporation of more electron-deficient selenium atom in place of sulfur atom. Here, strengthening the acceptor group diminishes the bandgap in D-A-D system.

In spite of combining the two aforementioned properties, the number of NIR II molecules are very rare. Many more strategies are essential for improving the emission wavelength to NIR II region. Combination of AIE (excellent intensity) and TICT effect (hypochromic shift in emission wavelength) is one of the best strategies for the designing of NIR II luminogens with high quantum yield (Fig. 1C). Controlling the twisted structure of a molecule in the aggregated state is sometimes tricky. For example, 2TT-*m*C6B was found to be ACQ in nature[45] (Fig. 2B). Although having molecules possessing TPA rotor-unit and hexyl chain, the ACQ behavior was because of coplanarity of thiophene-BBTD-thiophene core that resulted in long-term intermolecular interaction in aggregated state. The change in the position of hexyl chain from meta- to ortho-position of thiophene unit with respect to BBTD core provided good AIE nature with TICT effect in 2TT-*o*C6B molecule. The reason for this was assigned to the presence of hexyl chain in the ortho-position to cause steric hindrance to BBTD as well as high dihedral angled twisted conformation, which enabled slight intramolecular motion, prohibited intermolecular interaction in the aggregated state and showed emission at NIR II region (1030 nm). Later on, the introduction of ethyl hexyl chain in place of hexyl chain in 2TT-*o*C26B expanded the dihedral angle due to higher bulkiness. The expansion in the dihedral angle caused more TICT in this probe as well as boosted its emission to more NIR II regions with high quantum yield.

2.2 AIE-RTP active molecules

Phosphorescence take place when excitons will spin flip from singlet excited state S_1 to triplet excited state T_1 and the triplets will undergo radiative transition to the ground singlet excited state (S_0) under the influence of ISC process. ISC is dependent on the electronic energy level and configuration of corresponding organic molecule including effective spin orbit coupling (SOC).[46] According to El-Sayed rule, ISC is strong between different type of orbital, i.e., triplet (n π^*) to singlet (π π^*) or singlet (n π^*) to triplet (π π^*).[47] The ISC is also stronger if there is n-orbital perpendicular to the π-orbital. The rate of phosphorescence depends upon hybrid triplet states (T_1^*). T_1^* is combination of αn and βn, hence phosphorescence property is tunable (Fig. 3). In addition to high ISC, SOC and lower value of ΔE_{ST}, the non-radiative relaxation of triplet excitons is also necessary, as phosphorescence is spin forbidden process, and hence lifetime is quite higher than fluorescence and has higher probability getting quenched by oxygen and

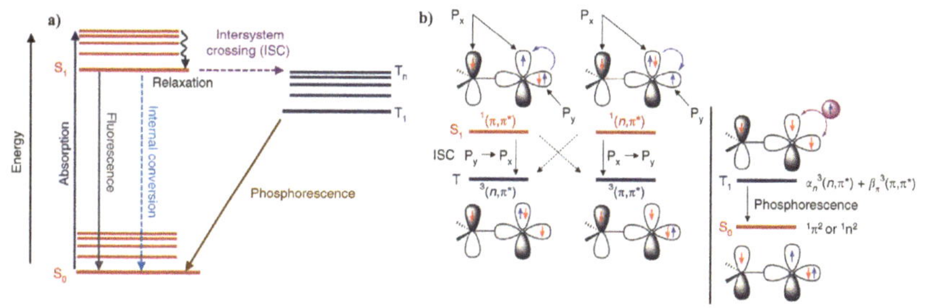

Fig. 3 (A) Jablonski diagram demonstrating different photophysical process occurred from singlet ground (S_0) to excited (S_1) state transitions like absorption, fluorescence, while exited triplet (T_1) to S_0 showed phosphorescence including intersystem crossing and internal conversion pathways (B) Schematic diagram explaining El-Sayed rule and molecular orbital hybridization of lowest triplet states for spin allowed transition. ISC is strong between $^1\pi$-π^* to ^{3}n-π^* or ^{1}n-π^* to $^3\pi$-π^*.

other environmental factors.[48,49] To construct pure organic RTP molecules many strategies have been developed including crystallization induced phosphorescence (CIP), directed halogen bonding, co-crystallization, H-aggregation, n-π^* transition, rigid and flexible host guest system, dopant based systems.[50–53]

To construct efficient RTP molecules two factors are mainly focused, one is to enhance singlet triplet intersystem crossing and the other is to minimize non-radiative decay from triplet state to ground state.[54] The very first approach toward this was introducing heavy halogen atom (Fig. 4A). The luminogens which contain carbonyl group interacts with the heavy halogen (generally bromine) and the lone pairs of oxygen resides in the empty orbitals of bromine, thereby enhancing the ISC which further intensifies the SOC process.[55] The other approach to generate good RTP materials through AIE systems is by minimizing non-radiative decay of triplet excitons. As the triplet excitons are very sensitive to environmental factors like oxygen, humidity, this needs to be a confined structure to reduce the decay which can be obtained by crystallization. This phenomenon is also known as crystallization induced phosphorescence (CIP).[56] It is reported that benzophenone and their derivatives retain intermolecular motion in crystalline state which enhances the probability of ISC from singlet to triplet excited state. Similarly, these relaxation pathways can be blocked by restricted intramolecular motion (RIM) introduced by crystalline lattice and also various intermolecular interactions could be realized from an AIE systems which extend the phenomenon of fluorescence to phosphorescence.[57,58] Crystallization of AIE active luminogens which has high ΔE_{ST} and n-π^*

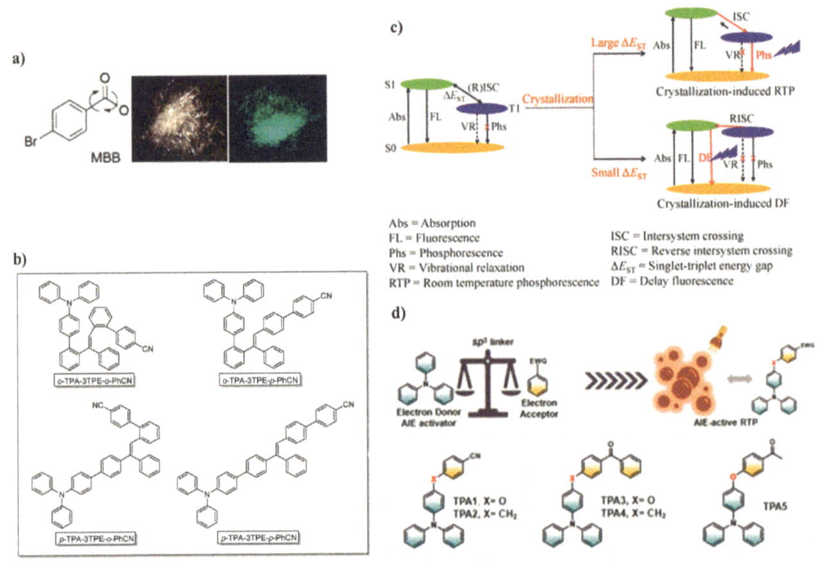

Fig. 4 (A and B) Halogen Bonding and Crystallization Induced Emission, Methyl 4-Bromobenzoate (MBB) structure and photograph under UV lamp Off and On. molecular structure arrangement chemical structure of TPA derivatives studied. (C) Schematic representation of ΔE_{ST} and crystallization based mechanism studied. (D) Chemical structure of TPA derivatives studied. Photo-physical studies and under UV ON and OFF of TPA derivatives.

transition with requisite molecular design can be a promising approach to design RTP active molecules from basic AIE luminogens. TPE (triphenylethene) based AIE active luminogen and incorporation of PhCN and TPA as electron acceptor and donor at different substitution positions resulted in huge steric hindrance and reasonable HOMO-LUMO including tuned ΔE_{ST} value (Fig. 4B and C). Introducing heavy halogen atom to organic chromophores hampered the PL efficiency yet this new design provided very efficient AIE-RTP molecules.[59]

A general design principle, which can be used to develop molecules having both AIE and RTP by developing sp^3 linker-D-A molecular structure has also been investigated. In this work intermolecular interactions were proposed to be more harmful for RTP molecules as they have longer lifetime than fluorescence, hence AIE mechanism was used to construct RTP molecules (Fig. 4D). TPA was used as an AIE activator connected through sp^3 linker with various donor groups. The linker enhanced the intersystem crossing leading to RTP lifetime efficiency. Five AIE active RTP molecules were synthesized with TPA as AIE property activator and various acceptors[56] and the role of linker and nitrogen atom of aromatic ring of TPA was investigated that made the complete structure twisted by about 110–120°. This twisted conformation renders to avoid π-π stacking and generate RTP in this system. All the intermolecular interactions minimize the molecular motion in TPA, which ultimately is responsible for AIE in TPA according to RIR principle. Hence, intramolecular CT between D-A and twisted conformation of the locally excited state of TPA are responsible for dual phosphorescence and AIE in a single D-A-D system with sp^3 linker.[60] Moreover, there are very few design strategies to develop AIE-RTP active molecules, nevertheless, very recently new design strategies are coming-up that may provide opportunities to develop such multifunctional molecules for many advanced applications.[61]

2.3 AIE-TADF active molecules

The design strategy of simultaneous AIE-TADF molecules is still under development. No rigid strategy is fixed for designing such materials and hence it is very difficult to propose any certain molecular design to realize these two properties simultaneously. Most importantly both the phenomenon comprise very different mechanisms since TADF must include both ISC and RISC.[62,63] However, before discussing the basic aspects of AIE-TADF materials it is auspicious to confer the basic aspects and molecular design strategy of common TADF molecules. The most important features of a TADF molecules are laying in the mechanism of TADF itself. When a molecule is excited, the singlet excitons are generated first and can go to the triplet state through ISC. To achieve TADF property, some of these triplet excitons should have the ability to up convert to singlet excitons via RISC. This RISC process is thermally activated, which is facilitated by the available temperature in the system during the process. Radiative decay of these singlet excitons, which results from the up conversion of triplet excitons gives a delayed fluorescence with thermal activation, hence termed as TADF.[64] The mechanism of TADF clearly indicates that the two important phenomena, ISC and most importantly RISC are needed to accomplish TADF. ISC is easily feasible in case of inorganic complexes because they comprise transition metals and heavy metals, which are very much susceptible to SOC (Fig. 5A). However, achieving sufficient ISC in a purely organic compound, which is completely metal free, is a challenging task due to the lesser viability of SOC.[65] A widely used tactic to overcome this difficulty is inclusion of non-metallic heavy atoms like bromine (Br), thioesters, carbonyl moiety etc. in the aromatic cores, that facilitate a better SOC, and thereby enhance the viability of ISC.[66] Additionally, robust

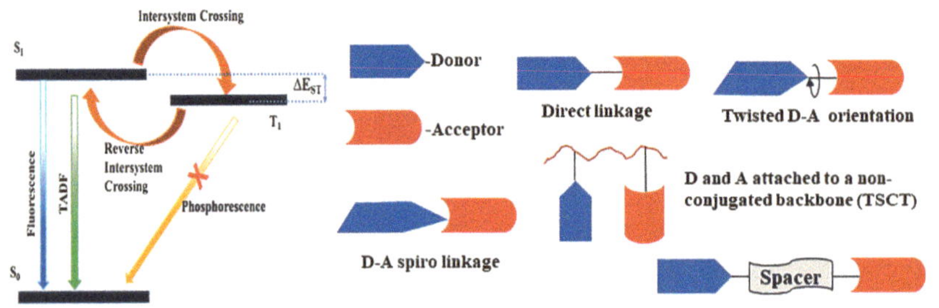

Fig. 5 (A) Jablonski representation of TADF mechanism. (B) Different donor-acceptor orientations in common TADF molecules.

intermolecular and intramolecular nonbonding interactions (H-bonding, π-π interaction etc.) also result in an enhancement in the ISC process.[67] Nevertheless, only realizing efficient ISC is not enough to accomplish TADF, yet the key process of up conversion of triplet excitons via RISC would still remain. The activation energy required to process the RISC is facilitated by the temperature available in the system that is the room temperature. A small energy difference between the triplet and singlet state (ΔE_{ST}) will make the process easier, whereas a large ΔE_{ST} value will offer a higher energy requirement, which overwhelms the available thermal energy in the system. If the system thermal energy cannot meet the energy requirement, that is the activation energy of RISC, then the whole setup for the delayed singlet emission will be a failure. Therefore, it is necessary to have a sufficiently small ΔE_{ST} value for thermal activation of RISC process, which should be generally less than 0.3 eV, although TADF molecules with ΔE_{ST} value of nearly up to 0.5 eV is reported.[68–70] The rate of RISC can be calculated by using the following formula-.

$$kRISC = \frac{k_F k_{TADF} \varphi_{TADF}}{k_{ISC} \varphi_F}$$

Here, k_{RISC}, k_F, k_{TADF}, and k_{ISC} represents rate constant of reverse intersystem crossing, fluorescence, TADF and intersystem crossing respectively and φ_F, φ_{TADF} represents quantum yields of fluorescence and TADF respectively.

The ΔE_{ST} value and the inherent photophysical properties associated with it are molecular properties and the molecular design plays an important role while achieving these properties. The most common approach to accomplish a small ΔE_{ST} value is donor-acceptor (D-A) molecular design, where an electronically rich molecular system (D) is linked to an electron deficient molecular system (A) (Fig. 5B).[71] This type of D-A molecular system provides very good separation of HOMO and LUMO of the molecule, where HOMO is mainly located towards the D-moiety and LUMO towards the A-moiety. The better separation of HOMO-LUMO results in small energy difference between the singlet state and the triplet state that enhances the probability of RISC process, thereby boosting the TADF quantum yield.[69]

3. Theranostic application

Recent development in medical technologies has helped overcome various life-threatening diseases. Theranostic application with NIR-AIE fluorescent materials has shown good efficacy owing to bright fluorescence with better specificity. Integration of therapeutics and diagnosis in a single platform is a potential approach in the biomedical field.[72] Two theranostic processes have achieved huge attention for emission guided specific ablation of cancer cell, photothermal therapy (PTT), and photodynamic therapy (PDT). The relaxation pathway of the fluorophores from the S_1 or T_1 to S_0 decides the diagnosis and therapeutics process. According to the Jablonski diagram, (Fig. 6A) absorption of excitation energy promotes an electron from S_0 to S_1. Relaxation of these excited electrons into its ground state follows several pathways- (1) fluorescence emission for relaxation from S_1 to S_0; (2) nonradiative dissipation through heat release; (3) recreation from T_1 to S_0 produces phosphorescence emission; (4) relaxation of the triplet electron through reactive oxygen species (ROS) production. Dissipated heat is employed for PTT and the generation of ROS can be utilized for PDT.

3.1 Photodynamic therapy (PDT)

PDT is one of the most powerful therapeutic methods in terms of apoptotic response towards cancer cells through ROS generation. Generation of ROS, that are deadly to the biological cells, regulates the efficiency of photosensitizer towards PDT applications. Synthesis of ROS follows two pathways- (1) reaction of triplet oxygen ($^3O_2$) with triplet electron and produce singlet oxygen ($^1O_2$). (2) Fenton like reaction of biomolecules with triplet electron which produce H_2O_2, $ONOO^-$, OH radical, etc. According to Fermi's Golden Rule, the SOC and ΔE_{ST} can modulate the rate of intersystem crossing.[73] Heavy atom incorporation in the photosensitizer (PS) is a well-known technique for more ISC. However, these

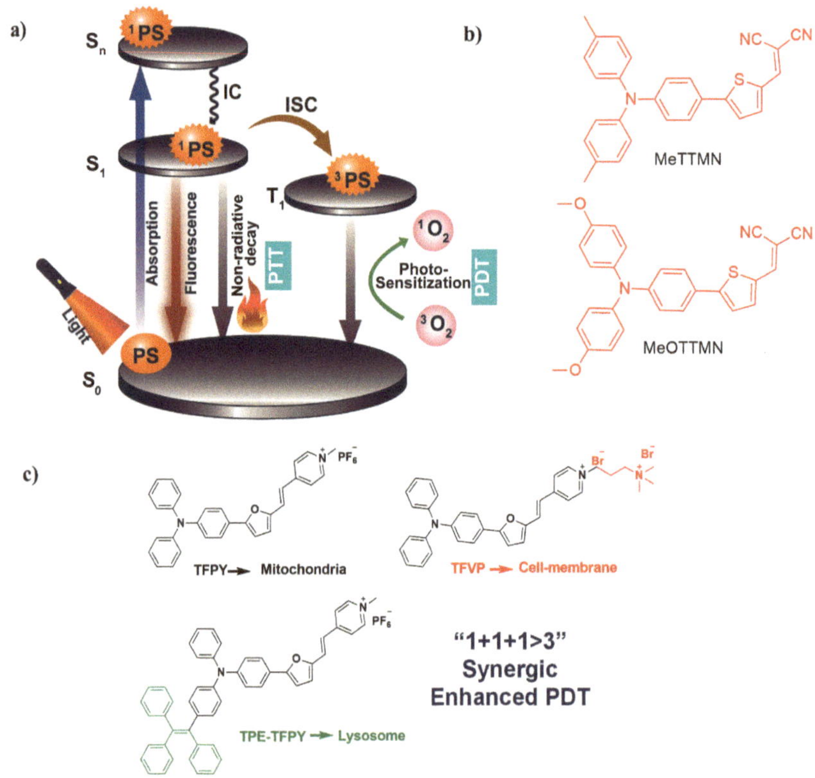

Fig. 6 (A) Pictorial presentation of a Jablonski diagram illustrates excitation of a photosensitizer (PS) and possible pathway of relaxation for further application toward PDT and PTT. (B) Chemical structure of MeTTMN and MeOTTMN. (C) demonstrates three different organelles targeting NIR probes used for enhanced PDT effect with combined application.

heavy atoms exhibit "dark toxicity," i.e., highly toxic to the cell without any light irradiation.[74] Hence, the heavy atom free PSs are much more desirable. Alleviation of the heavy atom by NIR-AIE PS not only provides less dark toxicity but also makes the synthesis process much easier. For NIR molecules, more π-conjugation gives small ΔE_{ST} between S_1 and T_1 states and better spin forbidden ISC as well as produce $^1O_2$. High repeatability, less invasiveness, low cost has made PDT much more superior alternative than the typical radiotherapy.[75,76]

The MeTTMN, and MeOTTMN probes (Fig. 6B), possess a combination of bulky rotor group (TPA) and unique D-A capability, showed decent AIE behavior with NIR emission,[44]. Further investigation of ROS generation ability of these two PS displayed excellent emission enhancement with

H2DCF-DA (ROS indicator) of up to 25 and 19 times concerning their initial emission respectively within short period of white light irradiation (90 s). Very less dark cytotoxicity of both the PS favored for their application to PDT and for the killing of cancerous cells. They applied both MeTTMN, and MeOTTMN photosensitizer to the HeLa cell (cancer cell) and checked the cell viability with increasing concentration under white light. For MeTTMN, 1×10^{-6} M concentration caused 50% cell death and further increment to 2.5×10^{-6} M concentration led to complete cell death. For MeOTTMN, 5×10^{-6} M concentration was required for overall cell apoptosis.

Specific organelle targeted PDT has opened a new era for long term tracing and therapeutics. Taking advantage of the organelle-specific application, three D-A based NIR emissive PS TFPy, TFVP, and TPE-TFPy (Fig. 6C) with the same backbone endowing different targeting specificity were developed.[77] These three probes showed good specific light-up and long-term imaging potential toward mitochondria, cellular membrane, and lysosome respectively inside the HeLa cell and 4T1 cancerous cell. All three possessed very low HOMO and LUMO bandgap of ~2 eV and had small ΔE_{ST} value of approximately 1 eV, which is sufficient for ROS generation. The long-term photostability and less toxicity under dark made all three good candidates for later therapeutic study. In the in vitro studies, the incubation of all "three in one" (1/3 of each PS in one pack) provided much better results (almost 100%) than the individual (less than 40%) to kill cancer cells under white light. From in vivo experiments, it was concluded that this novel strategy of combining organelle-specific PS exhibited superior therapeutic efficacy toward cancer cells killing than single PS.

3.2 Photothermal theranostics (PTT)

Photothermal therapy (PTT) based theranostics is another significant treatment modality that has a great impact on modern tumor therapy in terms of PTT and photoacoustic (PA) imaging.[78,79] It is an extended version of PDT, where generally, a NIR laser source is used for the excitation of the PS. The working principles of both PA and PTT rely on the heat release for the relaxation of the PS. In PA imaging, released heat leads to ultrasound emission wavelength and provides depth imaging to centimeter scale.[80] In PTT, the emitted heat is directly employed to ablate cancer cells. However, balancing between the radiative decay (fluorescence emission) and non-radiative decay (heat) is a crucial factor in gaining an efficient image-guided

PTT system. For AIE probes, restriction in intramolecular rotation leads to blocking of nonradiative decay and facilitates bright emission in the aggregate phase. On the contrary, the AIE molecules released heat for the rotation with no emission in solution state elevates the PTT effect. Also, in biological medium, more intramolecular interaction in D-A based coplanar NIR molecules has blocked all heat release pathway.[81] Therefore, controlling the molecular motion in the aggregated state is the only way to balance both heat generation and emission for extended accuracy in diagnosis and depletion of the tumor cell.

This hard challenge was successfully overcome by introducing the lipophilic BPBBT probe, (Fig. 7A) which exhibited albumin tailored NIR emission endowing AIE and TICT in equilibrium.[82] In photophysical studies, this D-A-D system possessed excellent TICT behavior with lowering of emission from low to high polarity solvent. Docking study of BPBBT inside human serum albumin (HSA) demonstrated the H-bonding interaction with Ser454 and Lys195. This interaction caused constrains in the BPBBT structure and formed large dihedral angled twisted conformation. Although the binding of HSA with BPBBT restricted intramolecular motion of the rotor group, the twisted structure also moved the equilibrium toward TICT state. TICT involvement facilitated BPBBT probe to achieve excellent photothermal behavior with little reduction in emission. The synthesized BPBBT nanoparticles (BPBBT NPs) in the presence of HSA was applied for in vivo studies in orthotopic CT26-Luc (colon cancered mice model). The imaging capability of this NIR II fluorophore was highly accurate for the metathesis visualization up to 0.5×0.3 mm along with primary tumor without any false signal. The PTT study showed that this probe could be able to generate 56 °C while the market available HAS/indocyanine green (ICG) (53 °C) also failed to reach that level. The optimal laser dose was also able to destroy whole cancer cells with very little damage to the adjacent normal cell. This NIR II fluorophore was able to cure colon cancer with high-resolution image guidance. Another D-A based NIR molecule was proposed for PTT application[83] and had a similar working principle, but the intermolecular motion was regulated by tailoring different chain lengths. For NIRb14, 2-decylmyristyl group attached to the thiophene unit produced better PA image and PTT effect than the shorter chain accompanied NIR6. Long and branched-chain hampered close and dense packing in the aggregated state because of the hydrophobic interaction. Bulky alkyl chain favored slight molecular motion owing to the much-twisted structure that introduced TICT effect as well as excellent PTT properties (Fig. 7B).

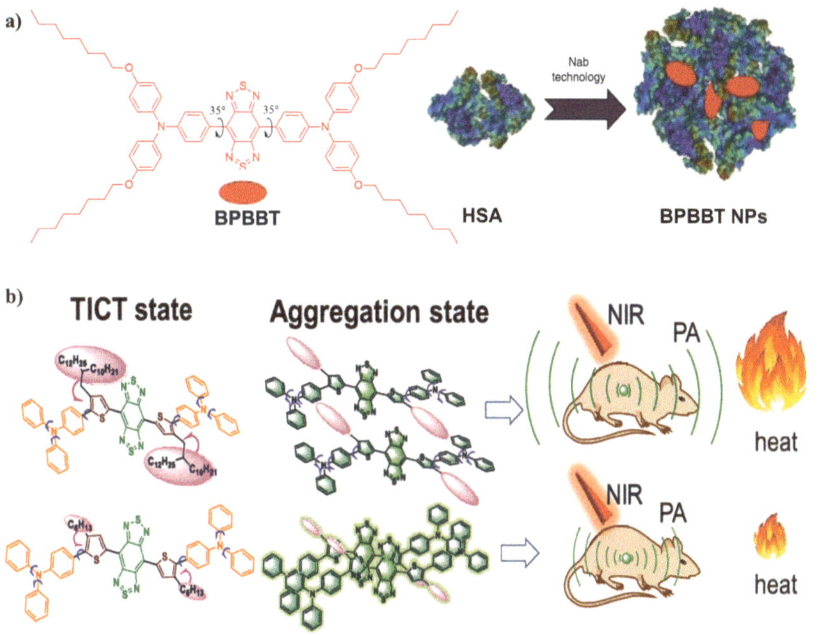

Fig. 7 (A) Schematic illustration of formation BPBBT NPs from BPBBT using Nab technology. (B) schematic illustration of the effect of bulky branched-chain on the intramolecular motion as well as in boosting PTT effect.

4. Fluorescence-based imaging with NIR-AIEgens

NIR emissive AIE luminogens exhibit potential imaging modality toward real-time visualization of specific biological structure and clinical application. Imaging with NIR I fluorophore has enticed considerable attention for its spatial resolution with less background fluorescence. In spite of high sensitivity of the NIR I probes, there are several limitations in terms of low depth imaging, interferential absorption and large photo scattering in critical tissue monitoring. Introduction of NIR II emissive molecule not only overcomes the restriction of NIR I but also provides unprecedented image clarity with excellent signal to noise ratio (S/N) for deep tissue monitoring.[22,84] More concern about high energy absorption toxicity has introduced two-photon absorption (2PA) probes leading to superficial imaging depth than one photon. In this method, same excitation energy is provided in terms of lower energy two photon having a sum of the same energy. Large cross-section area (σ) of the NIR probe makes them eligible for 2PA applications which shifts their excitation wavelength to more NIR window.[37,85] However, low survival rated operation demands further non-invasive methods in practical use. After a long effort, scientists have invented three-photon fluorescence (3PF) microscopy imaging NIR emissive organic molecules that are capable of overcoming all the drawbacks in real-world application.[86,87] Herein, the application of advanced NIR emissive AIEgens in wide range imaging, including brain inflammation, tumor monitoring, organelles specific imaging, etc. will be discussed.

4.1 Imaging of brain tumor and inflammation

Inflammation is one of the symptoms of several serious illnesses such as cardiac disease, cancer etc.[88] Severe inflammation causes carcinogenesis and leads to tumor growth.[89] Inflammation diagnosis inside the brain with the NIR probe faced severe obstacles form the protective blood-brain barrier (BBB), which will not allow probes to flow easily toward central nervous system. Hence, non-invasive recognition of the inflammation is highly desirable for the diagnosis in the premature stage. Recently, 2TT-*o*C6B probe was established for the visualization of the brain inflammation inside mice model[43] (Fig. 8A). This probe displayed absorption (808 nm) and emission (1030 nm) in the NIR region owing to TICT along with AIE behavior in the aggregated state. 2TT-*o*C6B was encapsulated with DSPE-PEG2000 and formed AIE dots (size 87 nm) to maintain the colloidal property and smooth circulation through blood. Neutrophils (NEs) are potent immune cell that have the capability to carry the NIR probes across BBB. After 1-h incubation of 2TT-*o*C6B AIE dots with NEs, good loading

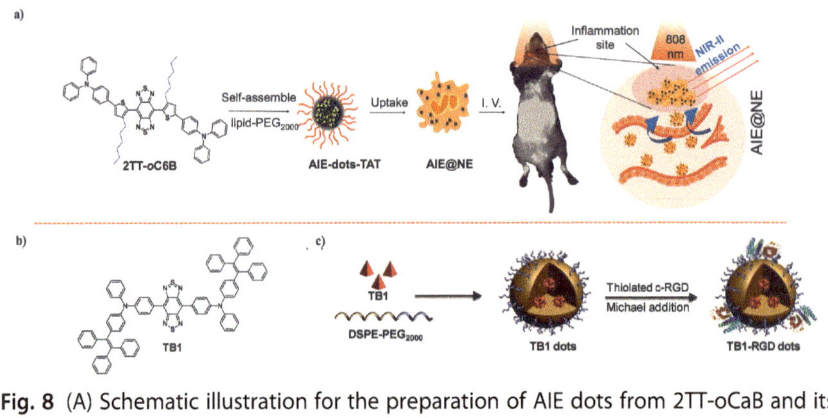

Fig. 8 (A) Schematic illustration for the preparation of AIE dots from 2TT-oCaB and its application for brain imaging. (B) chemical structure of TB1 AIEgen. (C) graphical representation of the TB1-RGD dots fabrication from TB1.

content of 1.4×10^{-5} nmol per cell resulted and AIE@NE presented brighter fluorescence emission than ICG@NE (commercially available indocyanine green with NEs). Motivated by the excellent loading capacity and high-intensity fluorescence emission, it was further injected into the brain inflamed mice model. After a 4 h post injection, the inflamed position was first detected and the emission intensity was gradually increased to give a transparent image after 12 h. In short, this small organic NIR AIEgen played an efficient role for the specific inflammatory site with signal to noise ratio of 30.6.

A new D-A conjugated AIEgen (**TB1**) for monitoring brain tumor was reported (Fig. 8B).[90] Anchoring of more number of bulky rotor units in donor *N,N*-diphenyl-4-(1,2,2-triphenylvinyl) aniline (**DPTPEA**) group enhanced the AIEgenic property of the probe and displayed high fluorescence quantum yield of 6.2% with emission maxima at 975 nm that extended up to NIR II region. They encapsulated **TB1** with DSPE-PEG2000 matrix to achieve well dispersed NPs in water. Integration of these dots with tumor targating cyclo(Arg-Gly-Asp-D-Phe-Lys(mpa))peptide (c-RGD) established **TB1**-RGD dots with specific tumor targeting ability (Fig. 8C). **TB1**-RGD dots were employed for further application in mice model having tumor inside the brain. After postinjection, cerebral blood vessels were visualized with superior clarity within 5 min. Up to 24 h time, the fluorescence intensity in the vessels decreased steadily and dots concentration accumulated in the tumor site. This result indicated that **TB1**-RGD dots were not only able to pass the BBB but also can image brain tumor with excellent sensitivity.

4.2 Image-guided tumor surgery

Fluorescence image directed tumor depletion has become a hot topic in the tumor treatment process. Fluorescence imaging technique is more sensitive compared to conventional imaging modality like positron emission tomography (PET), ultrasound imaging (USI), magnetic resonance imaging (MRI).[91] Further, an organic NIR molecule has made it more potent, owing to less cytotoxicity and profound penetration ability for the discovery of small or hidden tumors for primary diagnosis. The first D-A-D engineered NIR AIEgen was reported for the image directed tiny tumor resection,[92] bearing strong BBTD acceptor which was anchored with alkoxy substituted donor TPE unit (Fig. 9A). It included a geometrical isomer (α-DTPEBBTD-Cx and β-DTPEBBTD-Cx, where x is the number of carbon atoms in the

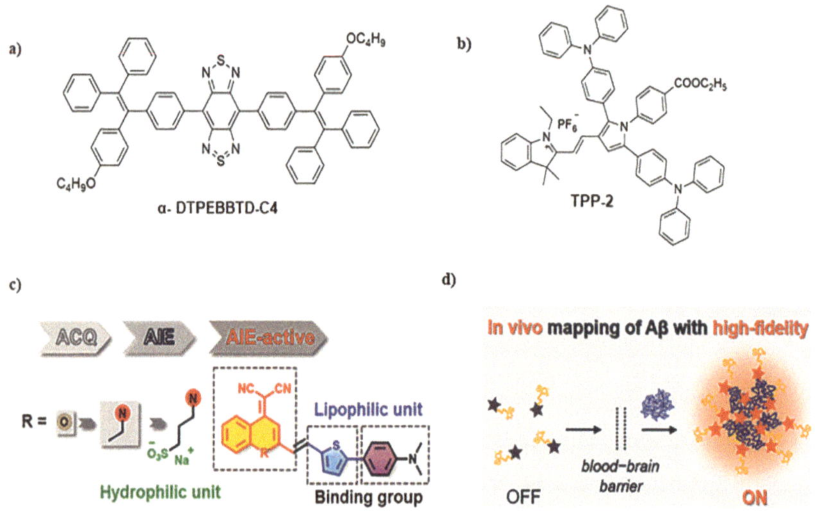

Fig. 9 (A) Chemical structure of α- DTPEBBTD-C4. (B) Chemical structure of TPP-2. (C) Schematic design strategy of NIR probe (QM-FN-SO3) through step by step structural modification. (D) Picturization of the "on-off" working principle of QM-FN-SO3 for Aβ mapping.

chain) introduced to control AIE features by varying the chain length and position. All the compounds were employed in their nanoparticle form (NPs) with DSPE-PEG2000 matrix for betterment in biocompatibility and colloidal property. Among all the NPs, α-DTPEBBTD-C4 NPs showed the highest fluorescence emission (801 nm), excellent photostability, which made it most eligible for in vivo application. To check cytotoxicity, the AIE NPs was injected inside healthy mice with almost twice the concentration than required for surgery application. It was noteworthy that high tumor-targeting efficacy of these bright fluorescent AIE NPs facilitated the resection of complete cancerous cells without harming any healthy tissue.

4.3 Organelle targeting wash free imaging

Living body are constructed with billions of cells with their specific functionality.[93] Cellular organelles like lysosomes, endoplasmic reticulum (ER) and mitochondria, etc., are the main building block of these cells. For a better understanding of biological complexion, it is important to know the individual role of the organelles. Limited number of tools have encouraged researchers to develop organelle targeting probe for visualization of morphology, size of specific organelle and also enables the study regarding the biochemical process. However, staining agents are designed depending upon mainly the membrane structures and their potential. For example, zwitterionic probes can stain the ER due to the electrostatic interaction with the enzyme present in its membrane and long alkyl chained compounds can target phospholipid bilayer for its hydrophobic nature.[94]

On the other hand, after probe staining, a post-wash process is always required to eliminate remaining free emissive particles. But this washing also reduces cell precision. Keeping all these things in mind, a triphenylpyrrole (TPP) derivative, TPP-2, was developed for wash free mitochondria imaging[95] (Fig. 9B). The presence of donor TPA group connecting with acceptor indolium group unveiled a large stokes shift of 219 nm. TPP-2 exhibited excellent AIE behavior under different DMSO/water and displayed emission maxima at 720 nm. This NIR AIEgen (TPP-2) was incubated with HeLa cells for in vivo imaging. Amphiphilic cationic TPP-2 was attracted towards negative potential mitochondrial membrane due to electrostatic interaction. It could target mitochondria within 10 s of incubation and showed exceptional photostability even after 6 h irradiation.

4.4 Imaging of Aβ plaque

Abnormal accumulation of misfolded proteins into amyloid β form results various neurodegenerative diseases like Alzheimer's disease (AD), Vascular dementia etc. AD, one of the most threatening complex neurodegenerative disorders which has already affected 44 million people worldwide, is still progressing severely mainly in the aging population.[96] This disease initiates slowly from synaptic dysfunction, which consequently lowers the memory functions along with task performing ability. Almost 40–70% of patients have accounted for dementia, and 90% over aged adults have been driven to death in the advanced stage AD.[97] The extracellular self-assembly of amyloid-β (Aβ) plaques leading to cross-β sheeted conformed fibrillar structure in the cerebral cortex of brain is considered to be a vital characteristic of this disease. After AD, vascular dementia is the second most threatening dementia resulting from Aβ aggregation in the cerebral arteries. This accumulation leads to lesser blood flow to the brain and severe damage to the vessels, which consequences sudden cut off of blood flow to some area of the brain. As per the previous reports, 10% people have witnessed dementia of mixed type both AD and Vascular dementia. But it was found that despite the homologous Aβ plaque formation in both dementias, the number of protein residue is different, 42 for AD and 39 for

Vascular dementia.[98] Commercial thioflavin-T and Congo red are the two most well-known staining agent generally used for Aβ fibrils.[99] But the lower emission wavelength (less than 550 nm) of those dyes compete with interferential autofluorescence of biological tissue. Despite this, limited BBB permeability and low depth imaging signal, severely restrict their application in the in vivo Aβ mapping.[100] Additionally, many AIE probes produce false signal due to the unexpected aggregation without binding Aβ. To address this dilemma, an ultrasensitive π-conjugated lipophilic AIE probe (QM-FN-SO3), which was non-emissive in biological medium and displayed instant NIR light up after binding with Aβ plaques was constructed.[101] (Fig. 9C and D). Step by step improvement of lipophilicity and hydrophilicity by introducing π-bridge thiophene unit and sulfonate group, respectively was followed. Tailoring of the sulfonate group position in the quinoline-malononitrile (QM) building block inversed its aggregation behavior and now water behaved as a good solvent toward QM-FN-SO3 and provided no emission. But due to aggregation, bright NIR emission (720 nm) was achieved in the higher ethanol volume fraction. The high binding affinity towards Aβ42 and very less background noise before binding made it a perfect contender for in vitro and in vivo mapping of Aβ plaques. The in vitro study of the probe inside the AD bearing mice brain slice provided amplified NIR signal with high fidelity in comparison to conventional Aβ mapping agent, ThS (thioflavin S). Further, QM-FN-SO3 probe was injected inside wild Alzheimer's free mice and APP/PS1 transgenic AD model. High NIR intensity inside the AD model was observed within 20 min, which concluded that QM-FN-SO3 could cross the BBB easily and could produce an ultra-fast signal for in vivo Aβ imaging.

5. AIE-RTP-active molecules for bio-imaging

Despite the above impressive results, more recently phosphorescent molecules have been reported for biological application due to the bright RTP emission, good photo-stability, long wavelength emission, good bio-compatibility, water dispensability, suitable size to cross membrane. In particular RTP molecules are attractive in biological imaging due to the longer lifetime of milliseconds to seconds which will discard the autofluorescence of several nanoseconds of the tissue. Most of the AIE-RTP active molecules reported are in crystalline form which makes it very scarce to use for biological imaging purpose because penetration inside cells is another major issue for the crystalline materials. Due to the fact that mechanism and synthesis protocols of AIE-RTP systems have not been investigated completely only few reports have been reported till date.[102]

Bio-imaging of Saponin, a natural, biocompatible plant-based glycosides, using AIE NPs, and phosphorescent nano-crystals (NCs) (Fig. 10) were reported recently.[103] These NPs were permeable through the membrane of Hela cells and imaging up to 100 ms. Alpha-hederine having unique cell permeabilization and membrane-lysis properties was used for this purpose. The deep red AIE-NPs were encapsulated using saponin which can stain cervical Hela cells very efficiently. The incubation time was 5 min which was much faster than any other conventional imaging methods. Very few deep red stating dyes are reported for clinical application and the best incubation time by those methods was 30 min, which shows the effectiveness of above method. This methodology was based on diffusion based ultrafast bio-imaging and shows promising results by such less incubation time overcoming endocytosis-based conventional bio-imaging process. In the same way, Saponin encapsulated NCs having AIE property, incubated in HeLa cells in 5 min gave bio-imaging up to 100 ms. This longer lifetime discards the auto-fluorescence and gave bio-imaging of millisecond ranges. Saponin encapsulated NCs also show very high photo-stability because of AIE property and provides a new path to AIE-RTP molecules for bio-imaging.

At the same time a unique in vivo real time phosphorescence bio-imaging was reported in living mice axillary lymph node for 10s by using top to down method to prepare water soluble organic NPs of phosphorescent luminogens (Fig. 11).[104] Notably, it was inert to biological

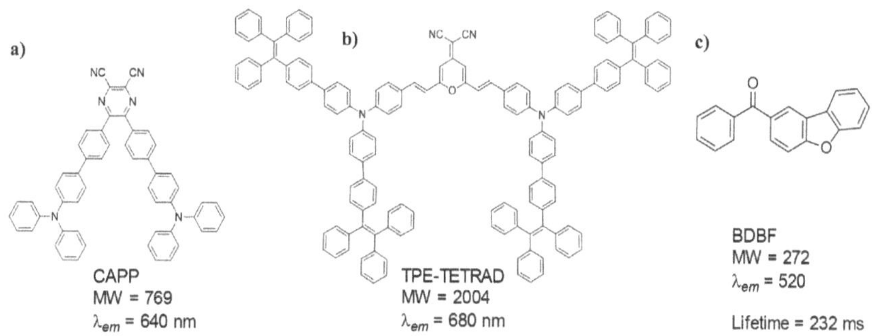

Fig. 10 Chemical structure and properties for (A) CAPP, (B) TPE-TETRAD, and (C) BDBF. CAPP and TPE-TETRAD are both cell-impermeable deep-red-emitting AIEgens. BDBF can form pure organic room-temperature phosphorescent nanocrystals.

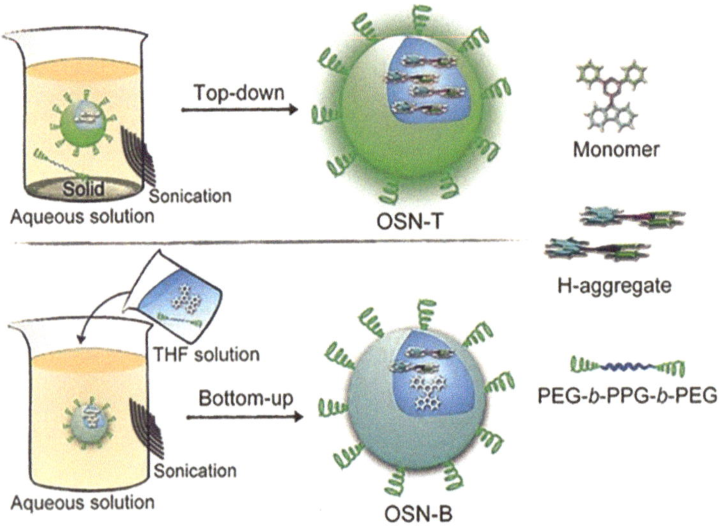

Fig. 11 Top to down and Bottom to up methods to synthesize OSNs-T and OSNs-B.

oxygen and water which could quench the triplet excitons. These NPs form aggregates in vivo because of strong intermolecular interaction and perhaps these aggregates are responsible for high photo-stability and imaging efficiency which further indicates the unrevealed potential application of AIE-RTP molecules in bio-imaging.

Another approach stretches that the RTP luminogens showed bio-imaging from visible to NIR region in difluoroboron β-diketonate including high uptake and bright imaging of HeLa cells including bio-compatibility (Fig. 12).[105] DFT study and spectroscopic findings showed that the emission was from the excited state dimers, and because of inter-molecular interaction and charge transfer in excited state the non-radiative decay reduced and also facilitated ISC, resulting in high brightness phosphorescent bio-imaging in aqueous media.

RTP crystals are impermeable to cell membrane and it is very difficult to make them permeable. A new strategy where nano-crystallization in (4-(4-(9H-carbazol-9-yl)butoxy)phenyl)(4-bromophenyl)methanone (C-C4-Br) system (Fig. 13) was reported. This method follows mainly three procedures; precipitation, seeding, followed by ultra-sonication. First nanoaggregates which are in amorphous were prepared by introducing luminogen in good solvent into its anti-solvent. These nano-aggregates were further subjected to direct crystallization seeding followed by ultra-sonication.

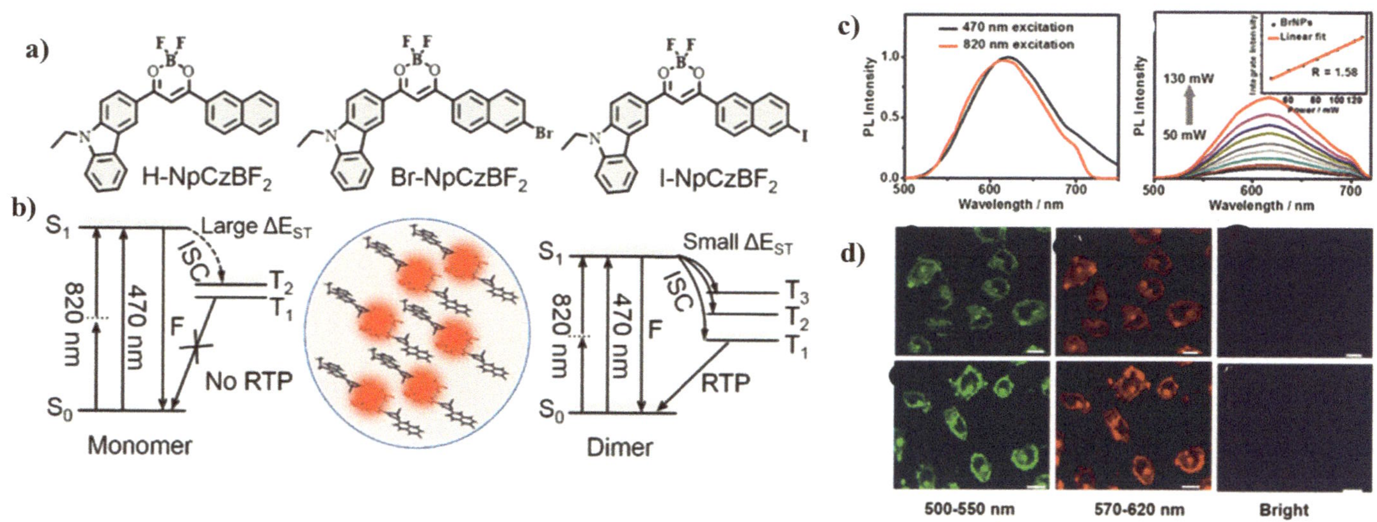

Fig. 12 (A) Chemical structure of the luminogen studied. (B) Mechanism elaborated through Jablonski diagram Photo-physical studies of AIE-RTP luminogens. (C and D) Imaging studies of HeLa cells treated with RTP NPs under Confocal laser microscopy.

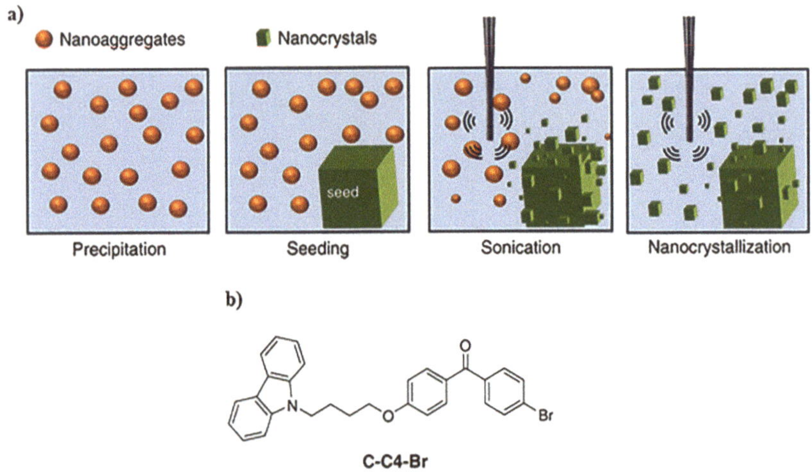

Fig. 13 (A) Schematic representation of new strategy of Nano-crystallization method to make RTP molecules for bio-imaging. (B) Chemical structure of persistent red RTP luminogen C-C4-Br (C) SEM images of C-C4-Br Nano-crystals.

Last step supports the nucleation and growth of nanocrystals. Size of the NCs were easily get tuned by changing solvent/anti-solvent ratio. These NCs of (C-C4-Br) showed lifetime of 0.14 s and excellent bio-compatibility along with no background auto-fluorescence and were further used for in vivo phosphorescence imaging of cancer cells.[106]

6. AIE-TADF emitter for biomedical applications

Phosphorescent probes exhibit lifetime values up to several milliseconds, which is sufficiently longer emission for optical imaging, yet their application in optical imaging is quite challenging due to relatively weak emission intensity, which originates from the triplet to singlet spin forbidden nature and quenching of triplet excitons in presence of oxygen.[107–109]

Such limitations hold back the growing demand for long-lived fluorescent probes with better brightness and efficient biocompatibility. In order to achieve optimal DF/PF ratios for fluorescence imaging, care should be taken in the selection of TADF emitters, mainly regarding the polarity effects on the TADF process. AIE-TADF luminogens are the most probable potential candidates for such applications, where TADF will provide longer lifetime with more fraction of exciton emission and AIE property under biological environment will provide robust nanoaggregates of the luminophores with enhanced emission intensity (Fig. 14). This growing interest on AIE-TADF

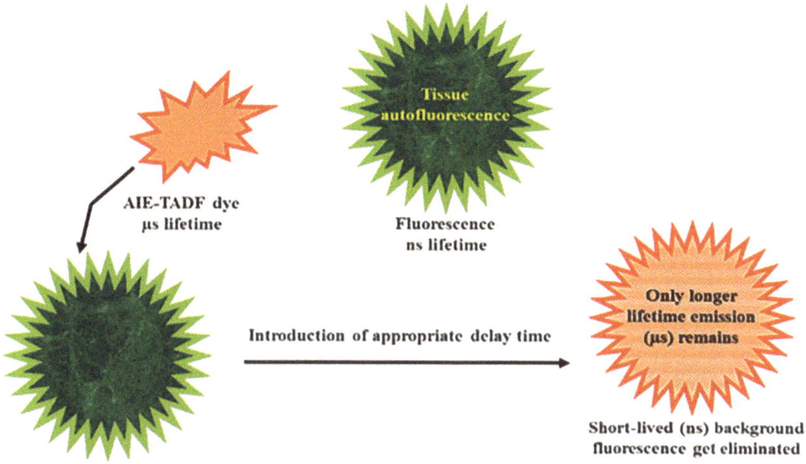

Fig. 14 Schematic representation of the strategy of using longer lifetime AIE-TADF dyes for TRFI.

emitters has encouraged practical applications of bioimaging. A new fluorescence organic dot (Odot) named Cpy-Odot was reported based on long lifetime green emissive TADF compiled with AIE, which exhibited 54.9% fluorescence quantum efficiency in film state with an average delayed lifetime of 9.3 μs.[33] The Odot-nanoprobes were fabricated effectively via DSPE-PEG2000 matrix encapsulation. Being small-sized, biocompatible, bright and robust with higher decay time in ambient condition, Cpy-Odots hold promising potential in confocal TRFI in living cells and their applicability was demonstrated in both zebrafish model and Hela cells after evaluating their cytotoxicity. Rather than nuclear region or in the cytoplasm of Hela cells, the Cpy-Odots displayed intense fluorescence signals on the plasma membrane (Fig. 15), which is an indication that these nanoprobes are capable of tracking plasma membranes dynamically.

Recently an unsymmetrical D-A-D type molecule namely M-1 comprised of N-(1H-indole-5-yl) acetamide and phenothiazine as two different donors and diphenylsulfone as the acceptor (Fig. 16) was reported.[110] The molecule exerted independent electronic transitions from the two different donors to the diphenylsulfone acceptor and performed dual TADF emission with a broader spectral wavelength distribution. The luminophore exhibited AIE with increasing water fraction in THF-H_2O mixture that reduces the contact probability between the emitter and oxygen, which can effectively prevent the triplet exciton quenching by oxygen. This benefit provided a

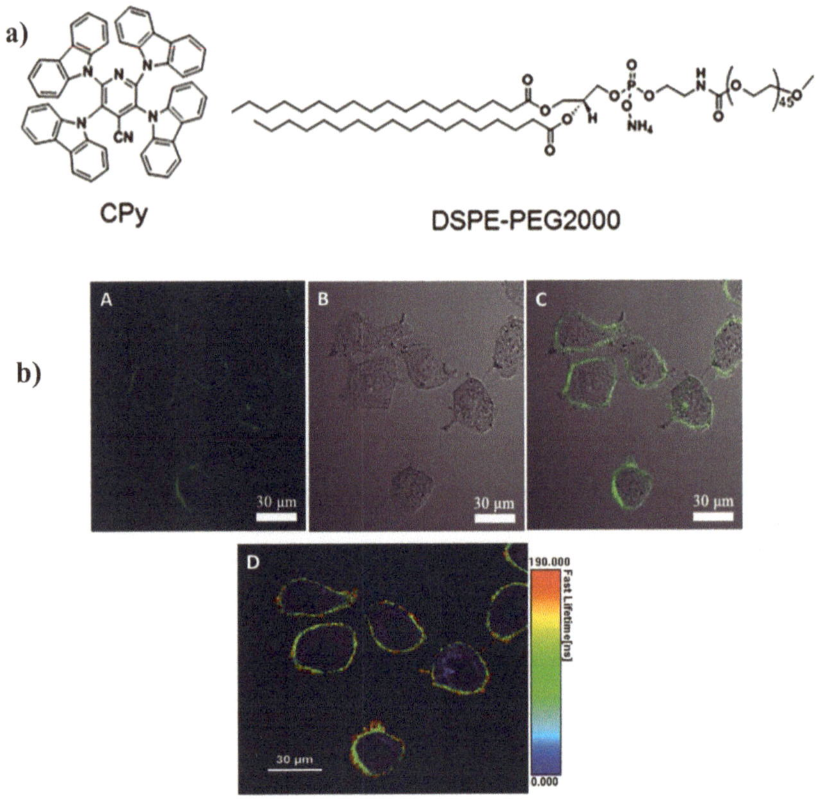

Fig. 15 (A) Molecular structure of CPy and DSPE-PEG2000. (B) Confocal TRFI in living cells in both zebrafish model and Hela cells.

superior time-resolved (TR) imaging competence with channel selectivity and an integrated and serialized intracellular local imaging evidence, as expected was observed by calibrating the two TR signals. This kind of dual emissive high lifetime signals can make it conceivable to avoid mutual energy transfers among luminogens and probable structural mismatching by allowing the two TADF signals with mutual complementarity throughout the probing.

A single component novel TADF amphiphilic emitter (AI-Cz-AM) was reported,[111] where the monomers were capable of forming self-assembly and performed potential application in confocal imaging and TRFI (Fig. 17). AI-Cz is an aromatic imide based TADF emitter with promising long emission lifetime, which exhibited excellent PL quantum yield in oxygen free nonpolar environments. Lipophilic TADF luminophore AI-Cz had a hydrophilic chain where the terminal is a positively charged molecule.

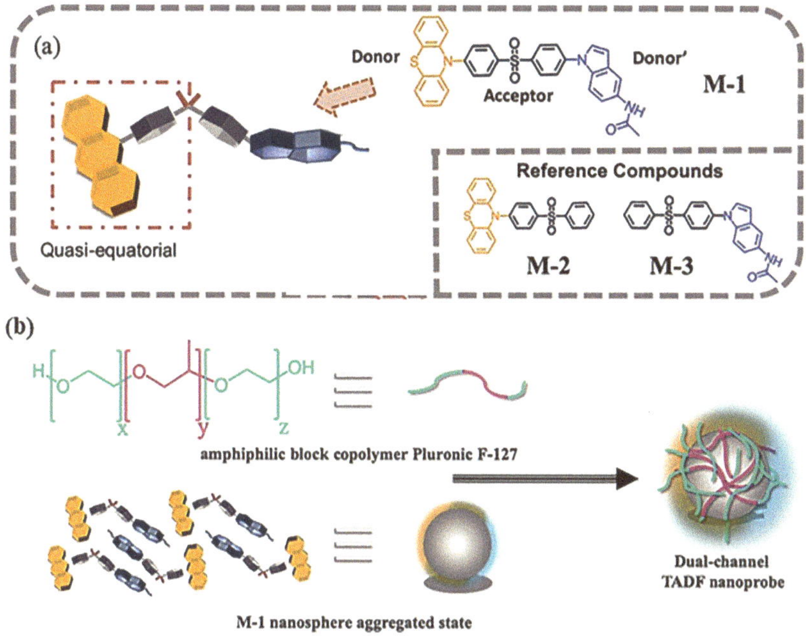

Fig. 16 (A) Molecular configurations of M_1 and reference molecules (M_2 and M_3). (B) Dual channel TADF nanoprobe prepared from aggregated nanosphere and block copolymer.

The amphiphilic nature of newly designed emitter (AI-Cz-AM) facilitates the spontaneous well defined self-assembly of the monomers in aqueous solution and results into a biocompatible and water soluble nanoprobe (AI-Cz-NP). The positively charged surface of the NPs endows enhanced cell penetration capability whereas the self-aggregation effectively limits the oxygen diffusion to the main emissive core (AI-Cz) that reduced the triplet quenching probability. The average delayed lifetime of the self-assembled AI-Cz-NP was found to be 10.68 μs in aqueous oxygenic solution. With the help of transmission electron microscopy (TEM) and CLSM the cellular uptake of the nanoprobe was examined and red fluorescence signals of AI-Cz-NP were observed, which designated that the nanoprobes could easily be taken up by the cells. The stable self-assembly nature of the probe was also revealed that maintained the assembled arrangements in HepG 2 cells. The material was further exploited in Fluorescence-lifetime imaging microscopy (FLIM) studies due to its promising cell labeling ability and TADF property, which displayed average long fluorescence lifetime of about 15 μs.

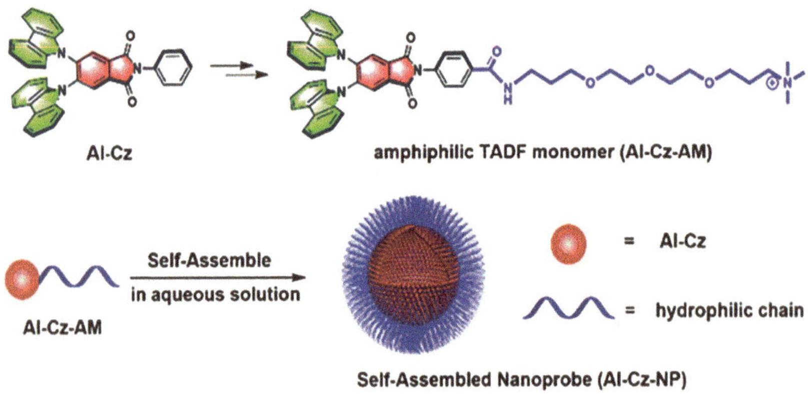

Fig. 17 Molecular structures and self-assembly of amphiphilic TADF monomer (Al-Cz-AM).

This longer emission time was found without any bio-oligomer encapsulation or further polymerization that explored the capability of self-assembly strategy to eliminate triplet quenching by oxygen.

A mitochondria targeting hydrophilic AIE-TADF (or AIDF) luminophore NID-TPP was also reported,[112] wherein NID is the main AIDF luminophore and TPP (TPP^{+}) was introduced to the core through a non-conjugated methylene linkage to bring hydrophilicity to the original hydrophobic core. NID-TPP displayed very weak emission without TADF feature in its single molecular state in aerated atmosphere and even in degassed environment.

The molecule exhibited AIDF feature with enhanced red emission in aerated environment that revealed suppression of the triplet state quenching by oxygen due to aggregation and this AIDF feature even in oxygen containing aerated atmosphere led to the potential use of the material in TRFI in living cells. The incorporation of TPP^{+} group offered hydrophilicity to the probe as well as endowed possible targeting and accumulation in mitochondria to achieve the expected AIE TADF and TRFI in living cells. Finally, TRFI and two photon luminescence imaging was performed in HeLa cells by using NID-TPP and TADF feature with red emission detected in mitochondria (further detected by co-staining with MitoTracker Green) with high selectivity within the quantified time domain with suppressed background noise (Fig. 18). These reports reveal that the well-designed AIDF molecules can be used as a promising tool for time resolved luminescence bio-imaging in oxygenated environments and organisms.

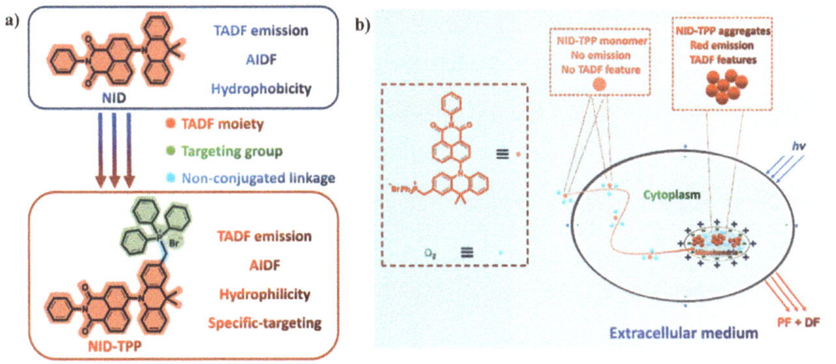

Fig. 18 (A) Molecular design of NID-TPP. (B) Proposed schematic mechanism of TRFI in HeLa cell mitochondria.

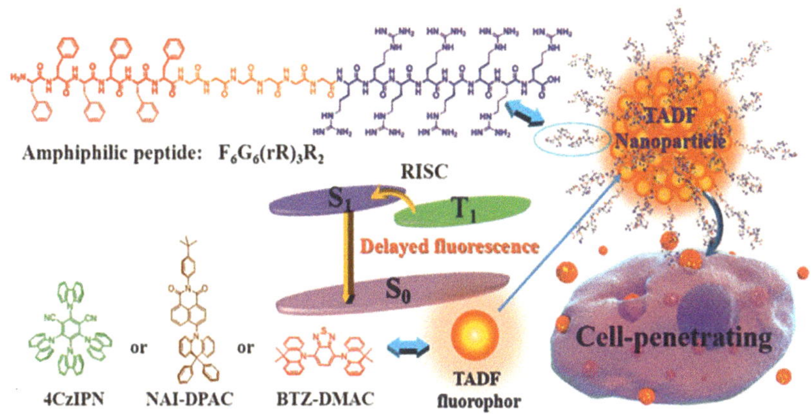

Fig. 19 TADF emitters with cell penetrating peptide of amphiphilic nature.

The first conventional TADF luminophores, which were capable of dispersing into NPs were formed by the self-assembly of a cell penetrating peptide [$F_6G_6(rR)_3R_2$] containing both hydrophobic and hydrophilic components (Fig. 19).[113] The three luminophores namely NAI-DPAC, 4-CzIPN and BTZ-DMAC, were dispersed into NPs and showed good cell penetrating ability and excellent biocompatibility (Fig. 19). With the aid of the peptide all the three TADF emitters exhibited long lived delayed emission (1.8–31.0/93.7 μs) in presence of air, offering them as a promising candidate for TRFI in living cells. The self-assembled emissive nature and the feasibility of long-lived emission in oxygen containing nature provide them with better biocompatibility, as a result in HeLa cells they exhibited emission lifetime up to 2 μs.

7. Conclusion and outlook

Since the coining of the concept of AIE in 2001, persistent efforts have created a library of AIEgens with their mechanistic understandings over last two decades. The real art of designing multifunctional AIE probe and their exploration in new applications has remained challenging but very exciting. However, the unique optical properties of AIE molecular probes and AIE-nanoparticles stimulated the establishment of excellent luminogenic platforms over conventional fluorescent materials for various biomedical applications. As compared to traditional fluorescence molecules, advanced AIE probes show high contrast imaging and deeper penetration depth, which have been demonstrated through NIR-I & NIR-II fluorescence imaging, room temperature after-glow imaging and photoacoustic imaging. AIE-NPs, AIE-NCs with tunable size and morphology for in vivo tumor imaging and tracking, vascular imaging, disease sensing, and image-guided surgery. In addition, through molecular engineering, AIEgens possessing photosensitization capability with the combination of new exciting photophysical process such as phosphorescence, delayed fluorescence offered an efficient sensing and imaging including precise image-guided photodynamic therapy, photothermal therapy at deep tumor tissues in living systems. This chapter highlights recent advancements of NIR-AIE molecular probes, AIE NPs, AIE-RTP and AIE-TADF luminogens, and their great potential for various imaging and theranostic applications, yet there are several challenges to be tackled for several practical applications. Despite these challenges, recent developments in AIE molecular probes, AIE-NPs, AIE-NCs light-up systems with improved optical properties, targeted therapy and low nonspecific toxicity, have been successfully utilized for practical trials such as disease diagnosis, cancer metastasis evaluation and drug screening, clearly indicating the bright future of the translational perspective of AIEgens.

References

1. Sinkeldam RW, Greco NJ, Tor Y. Fluorescent analogs of biomolecular building blocks: design, properties, and applications. *Chem Rev*. 2010;110:2579–2619.
2. Lin VS, Chen W, Xian M, Chang CJ. Chemical probes for molecular imaging and detection of hydrogen sulfide and reactive sulfur species in biological systems. *Chem Soc Rev*. 2015;44:4596–4618.
3. Weissleder R, Ntziachristos V. Shedding light onto live molecular targets. *Nat Med*. 2003;9:123–128.

4. Qian J, Tang BZ. AIE luminogens for bioimaging and theranostics: from organelles to animals. *Chem.* 2017;3:56–91.
5. Kobayashi H, Ogawa M, Alford R, Choyke PL, Urano Y. New strategies for fluorescent probe design in medical diagnostic imaging. *Chem Rev.* 2010;110:2620–2640.
6. Zhang T, Zhu X, Cheng CC, et al. Water-soluble mitochondria-specific ytterbium complex with impressive NIR emission. *J Am Chem Soc.* 2011;133:20120–20122.
7. D'Aléo A, Bourdolle A, Brustlein S, et al. Ytterbium-based bioprobes for near-infrared two-photon scanning laser microscopy imaging. *Angew Chem Int Ed.* 2012;51: 6622–6625.
8. Zhang Z, Fang X, Liu Z, et al. Semiconducting polymer dots with dual-enhanced NIR-IIa fluorescence for through-skull mouse-brain imaging. *Angew Chem Int Ed.* 2020;59:3691–3698.
9. Wegner KD, Hildebrandt N. Quantum dots: bright and versatile in vitro and in vivo fluorescence imaging biosensors. *Chem Soc Rev.* 2015;44:4792–4834.
10. Jing L, Kershaw SV, Li Y, et al. Aqueous based semiconductor nanocrystals. *Chem Rev.* 2016;116:10623–10730.
11. Li W, Dong K, Ren J, Qu X. A β-lactamase-imprinted responsive hydrogel for the treatment of antibiotic-resistant bacteria. *Angew Chem.* 2016;128:8181–8185.
12. Yin C, Zhen X, Fan Q, Huang W, Pu K. Degradable semiconducting oligomer amphiphile for ratiometric photoacoustic imaging of hypochlorite. *ACS Nano.* 2017;11:4174–4182.
13. Liu F, Tang C, Chen Q-Q, et al. Supramolecular π-π stacking pyrene-functioned fluorenes: toward efficient solution-processable small molecule blue and white organic light emitting diodes. *J Phys Chem C.* 2009;113:4641–4647.
14. Luo J, Xie Z, Lam JWY, et al. Aggregation-induced emission of 1-methyl-1,2,3,4,5-pentaphenylsilole. *Chem Commun.* 2001;1740–1741.
15. Deng C, Niu Y, Peng Q, Qin A, Shuai Z, Tang BZ. Theoretical study of radiative and non-radiative decay processes in pyrazine derivatives. *J Chem Phys.* 2011;135, 014304.
16. Liang J, Tang BZ, Liu B. Specific light-up bioprobes based on AIEgen conjugates. *Chem Soc Rev.* 2015;44:2798–2811.
17. Zhang X, Zhang X, Tao L, Chi Z, Xu J, Wei Y. Aggregation induced emission-based fluorescent nanoparticles: fabrication methodologies and biomedical applications. *J Mater Chem B.* 2014;2:4398–4414.
18. Forni A, Lucenti E, Botta C, Cariati E. Metal free room temperature phosphorescence from molecular self-interactions in the solid state. *J Mater Chem C.* 2018;6:4603–4626.
19. Yang Z, Mao Z, Xie Z, et al. Recent advances in organic thermally activated delayed fluorescence materials. *Chem Soc Rev.* 2017;46:915–1016.
20. Guibert EE, Petrenko AY, Balaban CL, Somov AY, Rodriguez JV, Fuller BJ. Organ preservation: current concepts and new strategies for the next decade. *Transfus Med Hemother.* 2011;38:125–142.
21. Tang Y, Li Y, Hu X, et al. "Dual lock-and-key"-controlled nanoprobes for ultrahigh specific fluorescence imaging in the second near-infrared window. *Adv Mater.* 2018;30:1801140.
22. Ding F, Zhan Y, Lu X, Sun Y. Recent advances in near-infrared II fluorophores for multifunctional biomedical imaging. *Chem Sci.* 2018;9:4370–4380.
23. Yang NJ, Hinner MJ. Getting across the cell membrane: an overview for small molecules, peptides, and proteins. *Methods Mol Biol.* 2015;1266:29–53.
24. Yu J, Diao X, Zhang X, et al. Water-dispersible, pH-stable and highly-luminescent organic dye nanoparticles with amplified emissions for in vitro and in vivo bioimaging. *Small.* 2014;10:1125–1132.
25. Deaton JC, Switalski SC, Kondakov DY, et al. E-type delayed fluorescence of a phosphine-supported Cu2(μ-NAr2)2 diamond core: harvesting singlet and triplet excitons in OLEDs. *J Am Chem Soc.* 2010;132:9499–9508.

26. Zhao Q, Huang C, Li F. Phosphorescent heavy-metal complexes for bioimaging. *Chem Soc Rev*. 2011;40:2508–2524.
27. Gorman A, Killoran J, O'Shea C, Kenna T, Gallagher WM, O'Shea DF. In vitro demonstration of the heavy-atom effect for photodynamic therapy. *J Am Chem Soc*. 2004;126:10619–10631.
28. Itoh T. Fluorescence and phosphorescence from higher excited states of organic molecules. *Chem Rev*. 2012;112:4541–4568.
29. Mukherjee S, Thilagar P. Recent advances in purely organic phosphorescent materials. *Chem Commun*. 2015;51:10988–11003.
30. Godumala M, Choi S, Cho MJ, Choi DH. Recent breakthroughs in thermally activated delayed fluorescence organic light emitting diodes containing non-doped emitting layers. *J Mater Chem C*. 2019;7:2172–2198.
31. Crucho CIC, Avó J, Diniz AM, et al. TADF dye-loaded nanoparticles for fluorescence live-cell imaging. *Front Chem*. 2020;8:404.
32. Shimizu M. Precious metal-free organic small molecule luminophores that exhibit room temperature phosphorescence. In: Tang Y, Tang BZ, eds. *Principles and Application of Aggregation-Induced Emission*. vol. 1. Springer International Publishing; 2019:43–76.
33. Ding D. Aggregation-induced emission luminogens for biomedical applications. In: Tang Y, Tang BZ, eds. *Principles and Application of Aggregation-Induced Emission*. vol. 1. Springer International Publishing; 2019:457–478.
34. Barman D, Gogoi R, Narang K, Iyer PK. Recent developments on multi-functional metal-free mechanochromic luminescence and thermally activated delayed fluorescence organic materials. *Front Chem*. 2020;8:483.
35. Li T, Yang D, Zhai L, et al. Thermally activated delayed fluorescence organic dots (TADF Odots) for time-resolved and confocal fluorescence imaging in living cells and in vivo. *Adv Sci*. 2017;4:1600166.
36. Genin E, Gao Z, Varela JA, et al. "Hyper-bright" near-infrared emitting fluorescent organic nanoparticles for single particle tracking. *Adv Mater*. 2014;26:2258–2261.
37. Gao Y, Feng G, Jiang T, et al. Biocompatible nanoparticles based on Diketo-Pyrrolo-pyrrole (DPP) with aggregation-induced red/NIR emission for in vivo two-photon fluorescence imaging. *Adv Funct Mater*. 2015;25:2857–2866.
38. Xiong X, Song F, Wang J, et al. Thermally activated delayed fluorescence of fluorescein derivative for time-resolved and confocal fluorescence imaging. *J Am Chem Soc*. 2014;136:9590–9597.
39. Weibel N, Charbonnière LJ, Guardigli M, Roda A, Ziessel R. Engineering of highly luminescent lanthanide tags suitable for protein labeling and time-resolved luminescence imaging. *J Am Chem Soc*. 2004;126:4888–4896.
40. Zhao Q, Zhang S, Liu Y, et al. Tetraphenylethenyl-modified perylene bisimide: aggregation-induced red emission, electrochemical properties and ordered microstructures. *J Mater Chem*. 2012;22:7387–7394.
41. Zhao Z, Su H, Zhang P, et al. Polyyne bridged AIE luminogens with red emission: design, synthesis, properties and applications. *J Mater Chem B*. 2017;5:1650–1657.
42. Li Y, Cai Z, Liu S, et al. Design of AIEgens for near-infrared IIb imaging through structural modulation at molecular and morphological levels. *Nat Commun*. 2020;11:1255.
43. Wang D, Su H, Kwok RT, et al. Facile synthesis of red/NIR AIE luminogens with simple structures, bright emissions, and high photostabilities, and their applications for specific imaging of lipid droplets and image-guided photodynamic therapy. *Adv Funct Mater*. 2017;27:1704039.
44. Wu W, Yang Y, Yang Y, et al. Molecular engineering of an organic NIR-II fluorophore with aggregation-induced emission characteristics for in vivo imaging. *Small*. 2019;15:1805549.

45. Liu S, Chen C, Li Y, et al. Constitutional isomerization enables bright NIR-II AIEgen for brain-inflammation imaging. *Adv Funct Mater*. 2020;30:1908125.
46. Zhao W, He Z, Lam Jacky WY, et al. Rational molecular design for achieving persistent and efficient pure organic room-temperature phosphorescence. *Chem*. 2016;1:592–602.
47. Lower SK, El-Sayed MA. The triplet state and molecular electronic processes in organic molecules. *Chem Rev*. 1966;66:199–241.
48. Celli JP, Spring BQ, Rizvi I, et al. Imaging and photodynamic therapy: mechanisms, monitoring, and optimization. *Chem Rev*. 2010;110:2795–2838.
49. Baryshnikov G, Minaev B, Ågren H. Theory and calculation of the phosphorescence phenomenon. *Chem Rev*. 2017;117:6500–6537.
50. Yuan WZ, Shen XY, Zhao H, et al. Crystallization-induced phosphorescence of pure organic luminogens at room temperature. *J Phys Chem C*. 2010;114:6090–6099.
51. Bolton O, Lee K, Kim H-J, Lin KY, Kim J. Activating efficient phosphorescence from purely organic materials by crystal design. *Nat Chem*. 2011;3:205–210.
52. An Z, Zheng C, Tao Y, et al. Stabilizing triplet excited states for ultralong organic phosphorescence. *Nat Mater*. 2015;14:685–690.
53. He Z, Zhao W, Lam JWY, et al. White light emission from a single organic molecule with dual phosphorescence at room temperature. *Nat Commun*. 2017;8:416.
54. Kenry CC, Liu B. Enhancing the performance of pure organic room-temperature phosphorescent luminophores. *Nat Commun*. 2019;10:2111.
55. Li J, Jiang Y, Cheng J, et al. Tuning the singlet–triplet energy gap of AIE luminogens: crystallization-induced room temperature phosphorescence and delay fluorescence, tunable temperature response, highly efficient non-doped organic light-emitting diodes. *Phys Chem Chem Phys*. 2015;17:1134–1141.
56. Schulman EM, Parker RT. Room temperature phosphorescence of organic compounds. The effects of moisture, oxygen, and the nature of the support-phosphor interaction. *J Phys Chem*. 1977;81:1932–1939.
57. Hirata S. Recent advances in materials with room-temperature phosphorescence: photophysics for triplet exciton stabilization. *Adv Opt Mater*. 2017;5:1700116.
58. Lee D, Bolton O, Kim BC, Youk JH, Takayama S, Kim J. Room temperature phosphorescence of metal-free organic materials in amorphous polymer matrices. *J Am Chem Soc*. 2013;135:6325–6329.
59. Kabe R, Adachi C. Organic long persistent luminescence. *Nature*. 2017;550:384–387.
60. Tao W, Zhubin H, Xiancheng N, et al. Thermochromic AIE dual phosphorescence via temperature-dependent sp3-linked donor-acceptor electronic coupling. *Nat Commun*. 2021;12:1364.
61. Feng G, Kwok RTK, Tang BZ, Liu B. Functionality and versatility of aggregation-induced emission luminogens. *Appl Phys Rev*. 2017;4, 021307.
62. Guo J, Li X-L, Nie H, et al. Robust luminescent materials with prominent aggregation-induced emission and thermally activated delayed fluorescence for high-performance organic light-emitting diodes. *Chem Mater*. 2017;29:3623–3631.
63. Araneda JF, Piers WE, Heyne B, Parvez M, McDonald R. High stokes shift anilido-pyridine boron difluoride dyes. *Angew Chem Int Ed*. 2011;50:12214–12217.
64. Huang B, Chen W-C, Li Z, et al. Manipulation of molecular aggregation states to realize polymorphism, AIE, MCL, and TADF in a single molecule. *Angew Chem Int Ed*. 2018;57:12473–12477.
65. Xu S, Liu T, Mu Y, et al. Molecule with asymmetric structure exhibiting aggregation-induced emission, delayed fluorescence, and mechanoluminescence. *Angew Chem Int Ed*. 2015;54:874–878.
66. Liang X, Tu Z-L, Zheng Y-X. Thermally activated delayed fluorescence materials: towards realization of high efficiency through strategic small molecular design. *Chem A Eur J*. 2019;25:5623–5642.

67. Yuan J, Wang Y, Li L, et al. Activating intersystem crossing and aggregation coupling by CN-substitution for efficient organic ultralong room temperature phosphorescence. *J Phys Chem C.* 2020;124:10129–10134.
68. Wong MY, Zysman-Colman E. Purely organic thermally activated delayed fluorescence materials for organic light-emitting diodes. *Adv Mater.* 2017;29:1605444.
69. Sun JW, Lee J-H, Moon C-K, Kim K-H, Shin H, Kim J-J. A fluorescent organic light-emitting diode with 30% external quantum efficiency. *Adv Mater.* 2014;26: 5684–5688.
70. Wei Q, Ge Z, Voit B. Thermally activated delayed fluorescent polymers: structures, properties, and applications in OLED devices. *Macromol Rapid Commun.* 2019;40: 1800570.
71. Parker CA, Hatchard CG. Triplet-singlet emission in fluid solutions. Phosphorescence of eosin. *Trans Faraday Soc.* 1961;57:1894–1904.
72. Zhu H, Fang Y, Miao Q, et al. Regulating near-infrared photodynamic properties of semiconducting polymer nanotheranostics for optimized cancer therapy. *ACS Nano.* 2017;11:8998–9009.
73. Ji S, Ge J, Escudero D, Wang Z, Zhao J, Jacquemin D. Molecular structure–intersystem crossing relationship of heavy-atom-free BODIPY triplet photosensitizers. *J Org Chem.* 2015;80:5958–5963.
74. Cakmak Y, Kolemen S, Duman S, et al. Designing excited states: theory-guided access to efficient photosensitizers for photodynamic action. *Angew Chem Int Ed.* 2011;50:11937–11941.
75. Fan W, Huang P, Chen X. Overcoming the Achilles' heel of photodynamic therapy. *Chem Soc Rev.* 2016;45:6488–6519.
76. Dolmans DE, Fukumura D, Jain RK. Photodynamic therapy for cancer. *Nat Rev Cancer.* 2003;3:380–387.
77. Xu W, Lee MM, Nie JJ, et al. Three-pronged attack by homologous far-red/NIR AIEgens to achieve 1+ 1+ 1> 3 synergistic enhanced photodynamic therapy. *Angew Chem Int Ed.* 2020;59:9610–9616.
78. Huang X, El-Sayed IH, Qian W, El-Sayed MA. Cancer cell imaging and photothermal therapy in the near-infrared region by using gold nanorods. *J Am Chem Soc.* 2006; 128:2115–2120.
79. Liu Y, Ai K, Lu L. Polydopamine and its derivative materials: synthesis and promising applications in energy, environmental, and biomedical fields. *Chem Rev.* 2014;114: 5057–5115.
80. Zhao Z, Chen C, Wu W, et al. Highly efficient photothermal nanoagent achieved by harvesting energy via excited-state intramolecular motion within nanoparticles. *Nat Commun.* 2019;10:1–11.
81. Li J, Rao J, Pu K. Recent progress on semiconducting polymer nanoparticles for molecular imaging and cancer phototherapy. *Biomaterials.* 2018;155:217–235.
82. Gao S, Wei G, Zhang S, et al. Lu W. albumin tailoring fluorescence and photothermal conversion effect of near-infrared-II fluorophore with aggregation-induced emission characteristics. *Nat Commun.* 2019;10:1–15.
83. Liu S, Zhou X, Zhang H, et al. Molecular motion in aggregates: manipulating tict for boosting photothermal theranostics. *J Am Chem Soc.* 2019;141:5359–5368.
84. Zhu S, Yung BC, Chandra S, Niu G, Antaris AL, Chen X. Near-infrared-II (NIR-II) bioimaging via off-peak NIR-I fluorescence emission. *Theranostics.* 2018;8:4141.
85. Wang S, Liu J, Goh CC, Ng LG, Liu B. NIR-II-excited Intravital two-photon microscopy distinguishes deep cerebral and tumor vasculatures with an ultrabright NIR-I AIE luminogen. *Adv Mater.* 2019;31:1904447.
86. Wang Y, Chen M, Alifu N, et al. Aggregation-induced emission luminogen with deep-red emission for through-skull three-photon fluorescence imaging of mouse. *ACS Nano.* 2017;11:10452–10461.

87. Qin W, Alifu N, Lam JW, et al. Facile synthesis of efficient luminogens with AIE features for three-photon fluorescence imaging of the brain through the intact skull. *Adv Mater.* 2020;32:2000364.
88. Song Z, Mao D, Sung SHP, et al. Activatable fluorescent nanoprobe with aggregation-induced emission characteristics for selective in vivo imaging of elevated Peroxynitrite generation. *Adv Mater.* 2016;28:7249.
89. Pasparakis M, Vandenabeele P. Necroptosis and its role in inflammation. *Nature.* 2015;517:311–320.
90. Sheng Z, Guo B, Hu D, et al. Bright aggregation-induced-emission dots for targeted synergetic NIR-II fluorescence and NIR-I photoacoustic imaging of orthotopic brain tumors. *Adv Mater.* 2018;30:1800766.
91. Willmann JK, Van Bruggen N, Dinkelborg LM, Gambhir SS. Molecular imaging in drug development. *Nat Rev Drug Discov.* 2008;7:591–607.
92. Liu J, Chen C, Ji S, et al. Long wavelength excitable near-infrared fluorescent nanoparticles with aggregation-induced emission characteristics for image-guided tumor resection. *Chem Sci.* 2017;8:2782–2789.
93. Marth JD. A unified vision of the building blocks of life. *Nat Cell Biol.* 2008;10:1015.
94. Alam P, He W, Leung NL, et al. Red AIE-active fluorescent probes with tunable organelle-specific targeting. *Adv Funct Mater.* 2020;30:1909268.
95. Ren F, Liu P, Gao Y, et al. Real time bioimaging for mitochondria by taking the aggregation process of aggregation-induced emission near-infrared dyes with wash-free staining. *Mater Chem Front.* 2019;3:57–63.
96. Mukherjee S, Madamsetty VS, Bhattacharya D, Roy Chowdhury S, Paul MK, Mukherjee A. Recent advancements of nanomedicine in neurodegenerative disorders theranostics. *Adv Funct Mater.* 2020;30:2003054.
97. Prince M. Alzheimer's disease facts and figures. *Alzheimers Dement.* 2018;14:367–429.
98. Prelli F, Castano E, Glenner GG, Frangione B. (1988), differences between vascular and plaque core amyloid in Alzheimer's disease. *J Neurochem.* 1988;51:648–651.
99. Hong Y, Meng L, Chen S, et al. Monitoring and inhibition of insulin fibrillation by a small organic fluorogen with aggregation-induced emission characteristics. *J Am Chem Soc.* 2012;134:1680–1689.
100. Dzwolak W, Pecul M. Chiral bias of amyloid fibrils revealed by the twisted conformation of Thioflavin T: an induced circular dichroism/DFT study. *FEBS Lett.* 2005;579:6601–6603.
101. Fu W, Yan C, Guo Z, et al. Rational design of near-infrared aggregation-induced-emission-active probes: In situ mapping of amyloid-β plaques with ultrasensitivity and high-fidelity. *J Am Chem Soc.* 2019;141:3171–3177.
102. Nicol A, Kwok RTK, Chen C, et al. Ultrafast delivery of aggregation-induced emission nanoparticles and pure organic phosphorescent nanocrystals by Saponin encapsulation. *J Am Chem Soc.* 2017;139:14792–14799.
103. Yang J, Zhen X, Wang B, et al. The influence of the molecular packing on the room temperature phosphorescence of purely organic luminogens. *Nat Commun.* 2018; 9:840.
104. Fateminia SMA, Mao Z, Xu S, Yang Z, Chi Z, Liu B. Organic nanocrystals with bright red persistent room-temperature phosphorescence for biological applications. *Angew Chem Int Ed.* 2017;56:12160–12164.
105. Wang X-F, Xiao H, Chen P-Z, et al. Room temperature phosphorescence from excited dimers in self-assembled nanoparticles under visible and near-infrared irradiation in water. *J Am Chem Soc.* 2019;141:5045–5050.
106. Jia X, Shao C, Bai X, et al. Photoexcitation-controlled self-recoverable molecular aggregation for flicker phosphorescence. *Proc Natl Acad Sci U S A.* 2019;116:4816.
107. Gesquiere AJ, Park S-J, Barbara PF. Hole-induced quenching of triplet and singlet excitons in conjugated polymers. *J Am Chem Soc.* 2005;127:9556–9560.

108. Notsuka N, Nakanotani H, Noda H, Goushi K, Adachi C. Observation of nonradiative deactivation behavior from singlet and triplet states of thermally activated delayed fluorescence emitters in solution. *J Phys Chem Lett.* 2020;11:562–566.
109. Charlesworth JM. Optical sensing of oxygen using phosphorescence quenching. *Sens Actuators B.* 1994;22:1–5.
110. Luo M, Li X, Ding L, et al. Integrating time-resolved imaging information by single-luminophore dual thermally activated delayed fluorescence. *Angew Chem Int Ed.* 2020;59:17018–17025.
111. Wei R, Zhang L, Xu S, Zhang Q, Qi Y, Hu H-Y. A single component self-assembled thermally activated delayed fluorescence nanoprobe. *Chem Commun.* 2020;56:2550–2553.
112. Ni F, Zhu Z, Tong X, et al. Hydrophilic, red-emitting, and thermally activated delayed fluorescence emitter for time-resolved luminescence imaging by mitochondrion-induced aggregation in living cells. *Adv Sci.* 2019;6:1801729.
113. Zhu Z, Tian D, Gao P, et al. Cell-penetrating peptides transport noncovalently linked thermally activated delayed fluorescence nanoparticles for time-resolved luminescence imaging. *J Am Chem Soc.* 2018;140:17484–17491.

CHAPTER FIVE

Aggregation-induced emission materials for protein fibrils imaging

Karma Patel, Syed Kabir Hussain Shah, and Panchami Prabhakaran*
School of Chemical Sciences, Central University of Gujarat, Gandhinagar, India
*Corresponding author: e-mail address: panchami.p@cug.ac.in

Contents

Abstract

Protein fibrillation is linked to many devastating diseases including neurodegenerative disorders. Fluorescence probes play a significant role in the detection of amyloid aggregates, monitoring amyloid kinetics, and in the development of amyloid inhibitors. Despite the considerable progress in this area, the mechanism of amyloid fibril formation in vivo is not completely understood. Recent studies in amyloidosis indicate that oligomers and prefibrillar species are more cytotoxic than the fibrils. Hence, early diagnosis of fibrillation has high therapeutical relevance. The gold standard for amyloid staining is thioflavin-T and its major drawbacks are aggregation caused quenching and inability in the detection of oligomers. New amyloid staining probes with novel properties are highly desirable as they can give valuable insights into the complicated process and can replace conventional probes. Aggregation-induced emission probes (AIE-probes) with desirable features are promising candidates in protein fibrils imaging.

Progress in Molecular Biology and Translational Science, Volume 185
ISSN 1877-1173
https://doi.org/10.1016/bs.pmbts.2021.06.011

AIE probes in staining different amyloid fibrils, monitoring amyloid kinetics, and early-stage conformers are reported. Other remarkable features are they can be modified as NIR probes, multifunctional probes, theranostic probes, and super-resolution imaging probes. We aim to provide a broad perspective on the progress attained with AIE probes in protein fibrils imaging and thereby emphasizing the scope of these smart probes in translative research.

1. Introduction

Misfolding and aggregation in peptides and proteins leading to fibrillation are associated with many debilitating diseases; includes common neurological disorders such as Alzheimer's and Parkinson's diseases and type II diabetes, rheumatoid arthritis, and cardiac arrhythmias, among others.[1,2] Proteins forming fibrillary, aggregated forms are termed amyloid fibrils and occurs through self-assembly, from the soluble form into insoluble unbranched fibrillar structures. The aggregation and accumulation of amyloid fibrils in organs and tissues are collectively known as amyloidosis. Experimental results suggest that it happens via a complex process.[3] In many amyloid diseases, the lack of apparent signs at the early condition creates challenges in detection and further treatment. Therefore, amyloids and their aggregation phenomenon have gained immense interest among chemists and biologists.[4,5] Many diagnostics on different classes are known in the literature that shed light on the kinetics of amyloid formation, detection, and development of amyloid inhibitors.[6] Scientific community has put enormous efforts into the design of materials as high sensitive probes in the detection of amyloid deposits that has strong therapeutical implications.[7,8] Despite the considerable endeavors from the scientists using modern chemical and analytical tools, understanding the driving forces and mechanism of amyloid fibril formation in vivo remains unclear. Considering the difficulties and importance of early diagnosis of amyloid formation, new diagnostic probes with novel properties are highly desirable as they can provide valuable insights into the amyloidosis process and will help in the design of potent inhibitors.[9] Hence, our purpose is to provide a broad perspective on the progress of research on aggregation-induced emission (AIE) probes in monitoring the protein fibrillation and diagnosis of amyloid deposits and thereby highlighting the scope of these smart probes in developing therapeutics to tackle the challenging diseases.

2. Protein fibrillation

Protein fibrillation is a process in which the misfolded proteins form amyloid fibrils. Various proteins and peptides serve as precursors for fibrillation and nearly 50 peptides and proteins are associated with amyloid diseases in human and domestic animals (Fig. 1A). Some of the techniques that help to study amyloidosis are magnetic resonance imaging (MRI), X-ray diffraction, circular dichroism, light scattering, Fourier transform infrared spectroscopy (FTIR), cryo-electron microscopy (cryo-EM), and atomic force microscopy (AFM).[10] With the help of modern techniques, scientists could reveal the structure of amyloid fibrils associated with a 11-residue fragment of protein transthyretin.[11] They adopt a canonical cross β-structure having a complex pattern. Recent developments include the structure determination of various amyloid fibrils at the atomic level.[12] Existence of multiple forms (polymorphism) is observed in small molecules[13,14] and proteins.[15] Multiple morphologies were observed in cryo-EM-based analysis on amyloid fibrils (Fig. 1B) taken from the brain with Alzheimer's disease (AD).[16] Interestingly, the structure of in vivo formed AD fibrils was different from the in vitro formed fibrils. The diversity found in amyloid structures may explain the difficulty in targeting the amyloid diseases.[17]

A three-step aggregation process in amyloidosis with sigmoidal curve kinetics involving nucleation, elongation, and equilibration phases is outlined in Fig. 2. Nucleation is a lag phase with the formation of oligomers.[18]

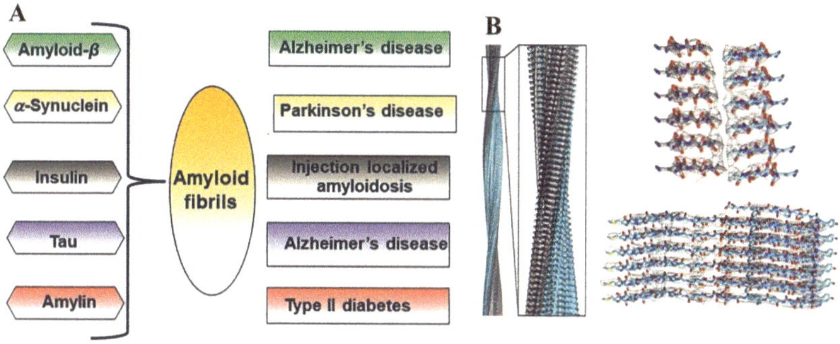

Fig. 1 (A) Few amyloid-forming proteins and the disease associated with them. (B) Cryo-EM structure of Aβ amyloid fibrils. *Panel (B) was taken from Kollmer M, Close W, Funk L, et al. Cryo-EM structure and polymorphism of Aβ amyloid fibrils purified from Alzheimer's brain tissue.* Nat Commun. *2019;10:4760.*

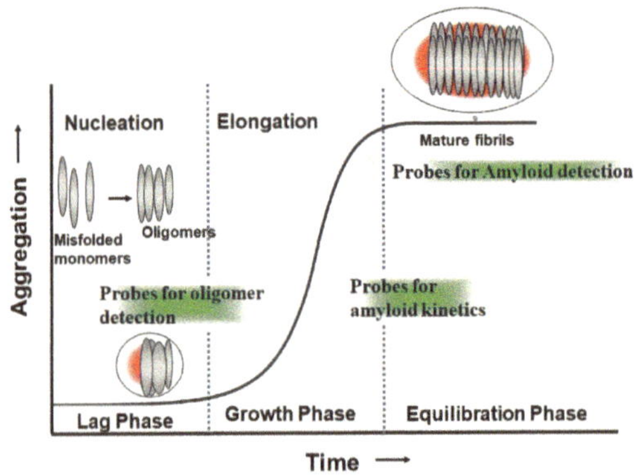

Fig. 2 General sigmoidal curve kinetics observed in protein fibrillation.

Until the end of the 1990s, amyloid fibrils were the main focus in developing therapeutics for amyloidosis. However, studies on various amyloid-forming proteins show that oligomers and prefibrillar species are more cytotoxic than the fibrils.[19] The findings led to the development of many diagnostics and therapeutics targeting the oligomers.[20]

3. Protein fibrils imaging

Among various bioimaging strategies, fluorescence spectroscopy has played a prominent role in studying the interaction of biomolecules and in the development of therapeutics.[21,22] Over the past years, realizing the scope of this non-invasive and cost-effective technique, considerable reports on fluorescence-based diagnostic probes for amyloid species are known.[6] They include organic fluorophores, inorganic quantum dots, nanoparticles, luminescent conjugated polythiophene (LCP), transition metal complexes, and fluorescent proteins.[23] Though organic fluorophores are widely used, their main drawback as aggregation caused quenching (ACQ) at high concentration has put a limitation on their use in quantitative analysis and bio-imaging experiments.[24,25]

4. AIE probes for amyloid detection

Molecules such as thioflavin T (ThT) and Congo Red (Fig. 3A) are the common staining probes in the study of protein aggregation.[26] The concept of aggregation-induced emission (AIE) paved the way for the development of AIE materials with a wide range of applications and among them,

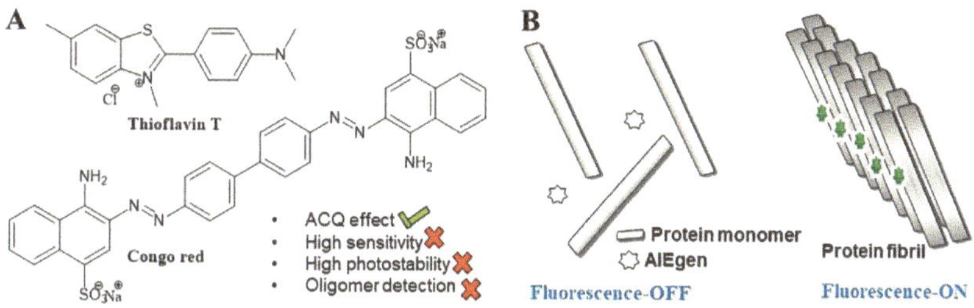

Fig. 3 (A) Chemical structures of commonly used fluorescent probes (ThT and Congo Red) in protein fibrils imaging. (B) Cartoon representation depicting AIEgen behavior with protein monomers and fibrils.

protein fibrils imaging is noteworthy.[8,27] Smart probes based on the AIE effect (AIEgens) have circumvented many limitations observed in conventional fluorophores such as the ACQ effect due to the π-π stacking at high concentrations, immiscibility in water due to the presence of many hydrophobic groups leading to aggregation in bio-environments, and poor photostability.[28,29] The AIE fluorogens are "switch off" (non-emissive) when dissolved and become "switch on" (emissive) when aggregate (Fig. 3B). The similarity in the aggregation events associated with the amyloid proteins and aggregation-induced emission process in AIE-probes make them appropriate candidates for the detection of amyloid aggregates.

Other advantages offered by the AIEgens are large stoke shift, high sensitivity and photostability, and remarkable signal-to-noise ratio. An ideal probe should have (i) high affinity and sensitivity, (ii) metabolic stability, (iii) low cell toxicity, and (iv) appropriate lipophilicity and blood-brain barrier (BBB) penetrability. The AIE probes can be tuned to possess these characteristics that are ideal for in vivo imaging of protein fibrils. Here, we have classified AIE probes for protein fibrils imaging based on the amyloid-forming proteins studied in the reported work. Though some of the probes may act as a generic probe for the amyloid deposits, this classification can be helpful for the readers to understand the progress in protein fibrils imaging with AIE fluorogens. The AIEgens mentioned here are given in Fig. 4.

4.1 Insulin

Insulin is an essential hormone that controls the metabolism of glucose in the human body and amyloid deposits have been diagnosed in patients suffering from type II diabetes.[30] The protein has two polypeptide chains,

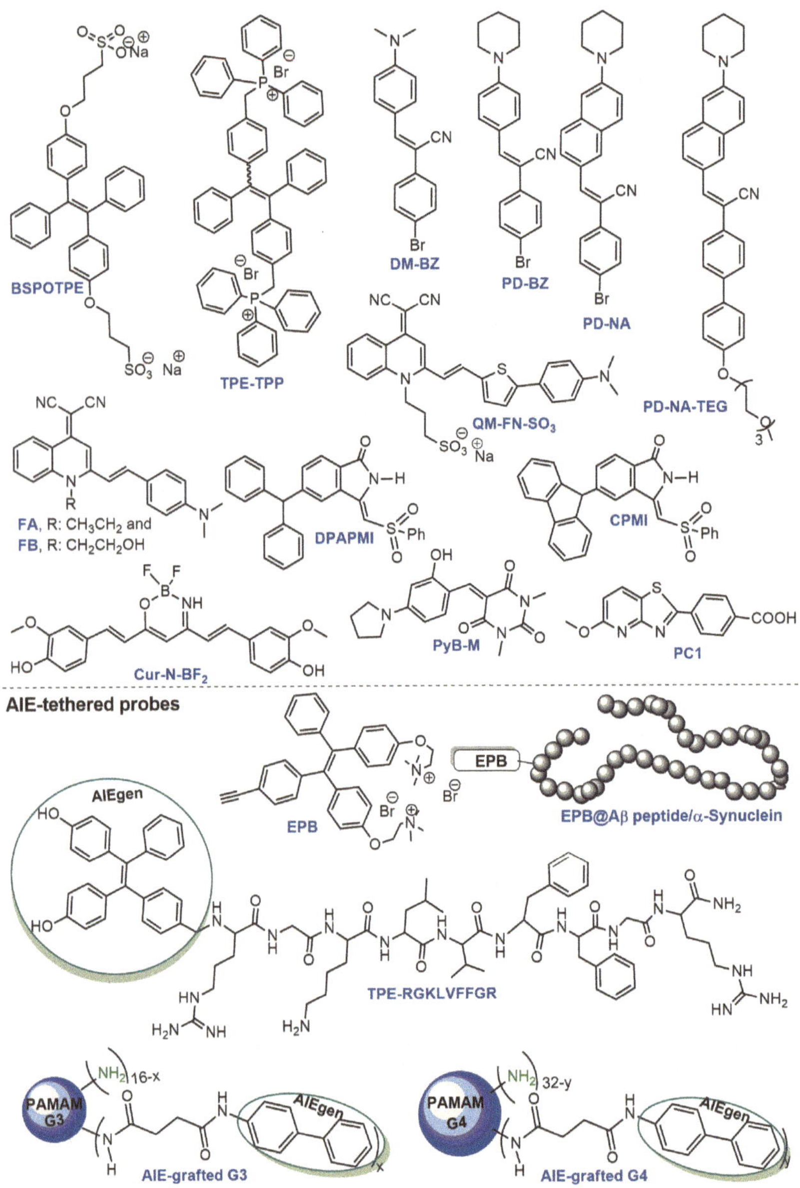

Fig. 4 Chemical structures of AIE materials in protein fibrils imaging.

chain A and B, mainly consists of helices among other secondary structures (A1–A8, A12–A19, and B9–B19, PDB: 4INS). It readily undergoes fibrillation at lower pH and high temperatures. In the proposed mechanism, the native monomer undergoes partial unfolding leading to nucleation and consecutive formation of fibrils that are rich in β-sheet structure.[31] As the protein has a high tendency to undergo fibrillation, it has been used as an ideal model to study the complicated amyloidosis process.

Tetraphenyl ethylene (TPE) based AIE probes are widely explored in the class of AIEgen due to the ease in synthesis and functionalization. Sodium 1,2-bis[4-(3-sulfonatopropoxyl)phenyl]-1,2-diphenylethene (BSPOTPE; Fig. 4), was studied in monitoring insulin amyloidosis.[32] It exhibited a distinct photophysical character, non-emissive in solution and with native form of insulin, and strong green light emission with fibrils of insulin. This indicated that the probe possesses AIE characteristics and can differentiate the fibrils and native form. Further, photoluminescence measurement studies supported the efficiency of the probe in detecting insulin fibrils and ex-situ monitoring fibrillation kinetics (Fig. 5A and B).

The data indicated that amyloidosis happened in three distinct steps: (i) nucleation with a faint emission in 30 min, (ii) elongation process with fluorescence enhancement in 1 h, and (iii) equilibration stage with no apparent change in the intensity. Further, it was used as an in situ inhibitor in insulin amyloidosis and the level of inhibition was proportional to the amount of BSPOTPE. SEM images of insulin taken after incubation with the fluorogen and control are given in Fig. 5C. Researchers demonstrated that the AIEgen molecule BSPOTPE displayed few important features, as a probe in the detection of insulin fibrils and monitoring the amyloid kinetics, and as an inhibitor in the amyloidosis.

Li and co-workers reported AIE grafted dendrimers in monitoring insulin fibrillation.[33] It is based on the concept that electrostatic interactions based binding affinity between a dendrimer and a protein can vary with changes in the size of the partners. Hence, based on the size effect, it is possible to distinguish monomers, oligomers, and fibrils with the help of fluorescence intensity. As a proof-of-concept, three and four generations of polyamidoamine dendrimer (PAMAM) grafted with 4-aminobiphenyl moiety (an AIEgen) were used. 4-Biphenylamide succinic acid was grafted to the third and fourth generations of dendrimers at different ratios to yield AIE-grafted G3 and AIE-grafted G4, respectively (Fig. 4). Under the experimental condition (pH 7.4), the negatively charged insulin (isoelectric point is 5.3) can interact with modified dendrimer. In both the series, AIE grafted dendrimers with the highest grafting ratio (G3-biph-4 and G4-biph-4) exhibited the strongest fluorescence with native insulin and insulin fibrils. Presence of the highest number of biphenyl units and hence more number on hydrophobic interactions between interacting partners are possible.

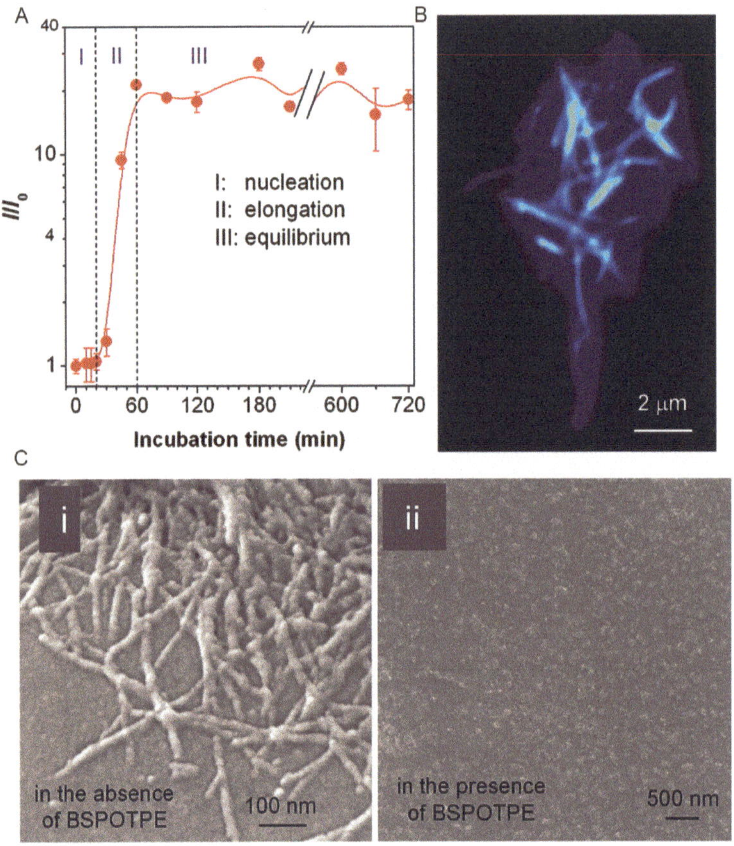

Fig. 5 (A) Insulin fibrillation kinetics monitored with BSPOTPE. (B) Staining of amyloid fibrils with BSPOTPE. (C) SEM images showing inhibition effect of the BSPOTPE on insulin fibrillation. *Reprinted with permission from Hong Y, Meng L, Chen S, et al. Monitoring and inhibition of insulin fibrillation by a small organic fluorogen with aggregation-induced emission characteristics.* J Am Chem Soc. *2012;134:1680–1689. Copyright 2012 American Chemical Society.*

Insulin fibrillation was monitored in situ with AIE grafted dendrimers in the presence of sodium chloride as it can affect electrostatic interactions if exist. A profound decrease in fluorescence intensity was observed in the interaction with native insulin at higher concentration of NaCl. Based on fluorescence data, it can be considered that interaction between AIE grafted dendrimers and native insulin is electrostatic, and hydrophobic with fibrils. With G3-biph-3/G4-biph-3, a decrease in fluorescence intensity with an increase in incubation time indicated the formation of fibrils. A pictorial representation to explain the fluorescence behaviors of G3-biph-3 and

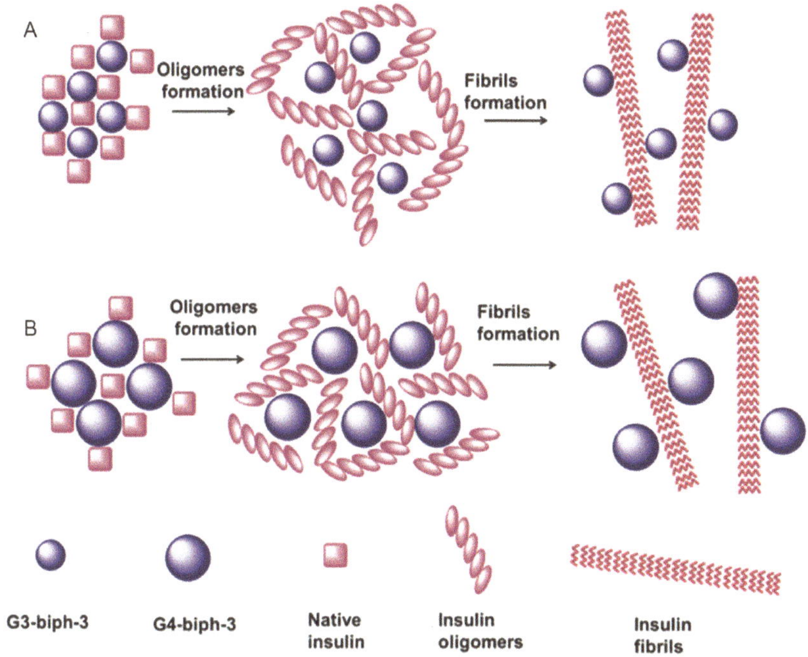

Fig. 6 Pictorial representation depicting the interaction of G3-biph-3 and G4-biph-4 with insulin monomers, oligomers, and fibrils. *Reprinted with permission from Huang Q, Xie J, Liu Y, Zhou A, Li J. Detecting the formation and transformation of oligomers during insulin fibrillation by a dendrimer conjugated with aggregation-induced emission molecule.* Bioconjug Chem. *2017;28:944–956. Copyright 2017 American Chemical Society.*

G4-biph-3 with the formation of oligomers and fibrils is outlined in Fig. 6. The insulin monomers are held by electrostatic interactions around the dendrimers restricting the rotation of biphenyl units and hence, results in the highest fluorescence intensity. When oligomers approach the surface of dendrimers, they can be repelled by their size and charge. They may tend to form a network as shown by middle representations in Fig. 6. More void space and hence more free movement of biphenyl units can occur in the molecular assembly formed between oligomers and G3-biph-3, when compared to the network formed between higher size G4-biph-3 dendrimers and oligomers. Thus, with proper design, functionalization, and size factors, AIE-tethered dendrimers can provide valuable information on dynamic protein fibrillation.

Most of the AIE probes fluorescence at the same wavelength in aggregated form and when associated with fibrils. This may create a false positive signal if the probe aggregates before binding with amyloid deposits.

To address this, two organic molecules with donor–acceptor-based charge transfer structure possessing AIE characteristics ((*Z*)-5-(diphenylamino)-3-((phenylsulfonyl)methylene)isoindolin-1-one (DPAPMI), (*Z*)-5-(9*H*-carbazol-9yl)-3-((phenylsulfonyl)methylene)isoindolin-1-one (CPMI), Fig. 4), were reported by Hazra and co-workers.[34] They formed spherical morphologies based on SEM images and gave aggregation-induced emission in buffer at 560 nm and 510 nm, respectively. They could stain amyloid fibrils (Fig. 7A) and exhibited a blue shift upon binding with fibrils. DPAPMI showed a wavelength shift (560–430 nm) with a color change from orange to blue, and 510 nm to 420 nm (green to blue) was observed with CPMI upon addition of insulin fibrils.

Staining probes known for amyloid imaging such as ThT do not affect the fibrillation kinetics. However, a different behavior was observed with these AIE dots on the insulin fibrillation process. They accelerated the nucleation and elongation stages, and it was supported by AFM study on the morphologies of fibrils grown at two different pH. The AFM images of fibrils taken from the incubation mixture with and without the probes at pH 2, indicated that fibrils were grown in the presence of the probe

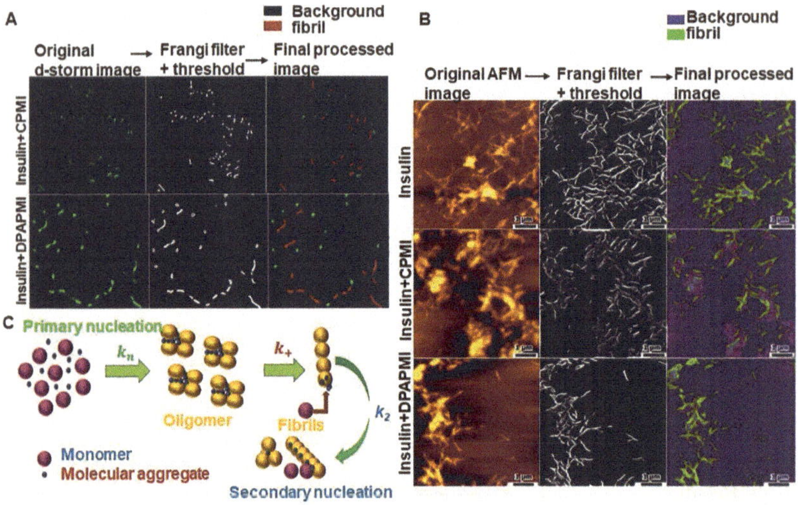

Fig. 7 (A) d-STORM super-resolution imaging of fibrils with CPMI and DPAPMI. (B) AFM images showing morphologies of fibrils at pH 2 with and without organic dots. (C) Pictorial representation of the effect of the probe in insulin fibrillation kinetics at pH 2. *Reproduced from Uddin A, Roy B, Jose GP, Hossain SS, Hazra P. Sensing and modulation of amyloid fibrils by photo-switchable organic dots.* Nanoscale. *2020;12:16805–16818, with permission from Royal Society of Chemistry.*

showed the formation of higher-order aggregates (Fig. 7B) with a high aspect ratio (length/thickness). A possible explanation for the elongation of fibrils in presence of organic dots is that the individual molecules bind with hydrophobic pockets in fibrils, allowing them to act as sites for further elongation (Fig. 7C). AFM images of the fibrils grown at room temperature for one day and by changing the pH to 7.4 showed a lower aspect ratio and a higher area fraction. It is already known that insulin fibrillation is destabilized at higher pH. Therefore, destabilization at the initial stage when moving to higher pH might have resulted in the formation of early aggregates. The organic probes on their surface may accelerate the secondary nucleation process leading to branching and thickening of fibrillar structures. These biocompatible probes accelerated the amyloid kinetics to form fibrils. Here, probes directed more toxic oligomers and prefibrillar species into less toxic fibrils. Such kinds of bioprobes may act as drugs in amyloidosis.

4.2 Lysozyme

One of the recent advancements in bioimaging is the use of near-infrared (NIR) probes. The advantages of NIR probes over the probes that emit visible light are a better signal-to-noise ratio and less absorption by the biological tissues. It can penetrate deeper and help us to image deeper structures with minimal photodamage to biological samples.[35] The commonly used strategy for fluorogens to exhibit long-wavelength emission is to incorporate flat and extended conjugation or with help of strong-donor acceptor skeletons which usually involves multi-step synthesis. Recent work done on fluorogens consists of barbituric acid as an electron-withdrawing group and aniline as an electron-donating group (for example, 1,3-dimethyl-5-(4-(pyrrolidin-1-yl)-benzylidene)pyrimidine-2,4,6(1H,3H,5H)-trione (PyB-M); Fig. 4) exhibited the AIE effect and were synthesized with a simple Knoevenagel reaction.[36] The dyes showed emission in the orange-to-red region with good quantum yield in solid-state. The researchers further demonstrated the use of the AIE probe in detecting fibrils formed from hen egg white lysosome (HEWL) as a model.

4.3 Amyloid-β

Among the chronic neurological disorders, Alzheimer's disease (AD) is observed mainly in 6% of the elderly population resulting in conditions such as memory disorder, executive dysfunction, etc. Research shows that a 39–42 amino acid peptide (amyloid-β) is associated with AD.[37]

The formation of plaques from amyloid-β (Aβ) is a pathological effect and is one of the biomarkers in AD.[38] The Aβ fibrils formed in the nerve cells of the brain tissues are released into the extracellular matrix after the neuronal death and forms the Aβ plaques. Early diagnosis of AD is difficult as the symptoms often match with signs associated with aging.

In a recent report, quinoline-based AIEgens were demonstrated to possess dual diagnostic properties (1-ethyl-2-(4-dimethylaminostyryl)-4-(dicyanomethylene)-1,4-dihydroquinoline (FA) and 1-(2-hydroxyethyl)-2-(4-dimethylaminostyryl)-4-(dicyanomethylene)-1,4-dihydroquinoline (FB); Fig. 4). They were used in the fluorescence imaging of amyloid-β plaques and lipid droplets.[39] Recent studies show that lipid droplet is an AD biomarker.[40] An increase in fluorescence intensity with a blue-shift was observed in the binding studies with Aβ_{1-42} aggregates, and a greater response was observed with FB. The standard MTT assay with FB on SH-SY5Y cells showed low cellular toxicity. Further, FB gave promising results on the fluorescence imaging of the Aβ aggregates and lipid droplets at the cellular level and in brain sections of APP/PS1 and APP/PS1/Tau transgenic mice (Fig. 8).

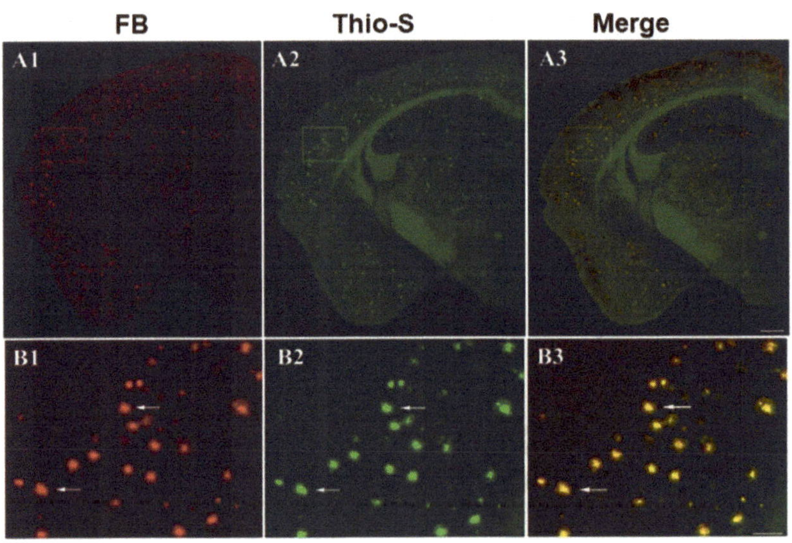

Fig. 8 Colocalization fluorescence imaging in the brain sections of APP/PS1 transgenic mouse with double labelling of FB (red) and thioflavin-S (green) and right is merged image. *Reproduced with permission from Wang Y, Qiu Y, Sun A, et al. Dual-functional AIE fluorescent probes for imaging β-amyloid plaques and lipid droplets.* Anal Chim Acta. *2020;1133:109–118. Copyright 2020 Elsevier.*

The work indicated that molecule FB can serve as one of the efficient probes for the diagnosis of amyloid deposits and lipid droplets.

Tetraphenylethylene (TPE) is a well-studied AIE moiety. A bioconjugate AIE probe consists of TPE component and a short peptide RGKLVFFGR (TPE-RGKLVFFGR; Fig. 4) to detect amyloid fibrillation was reported by Pradhan *et al.*[41] Herein, the selectivity of the peptide to amyloid protein/peptide and light-up fluorescence by the AIEgen moiety upon binding with amyloid deposits are combined in the smart design of the diagnostic probe. The TPE-peptide probe was then analyzed to detect three different fibrils: amyloid-β fibrils associated with AD, insulin fibrils found in type II diabetes patients, and lysozyme fibrils. The fluorescence switch-*on* happened in the presence of all the three fibrils and performance was better with Aβ-fibrils (Fig. 9A and B).

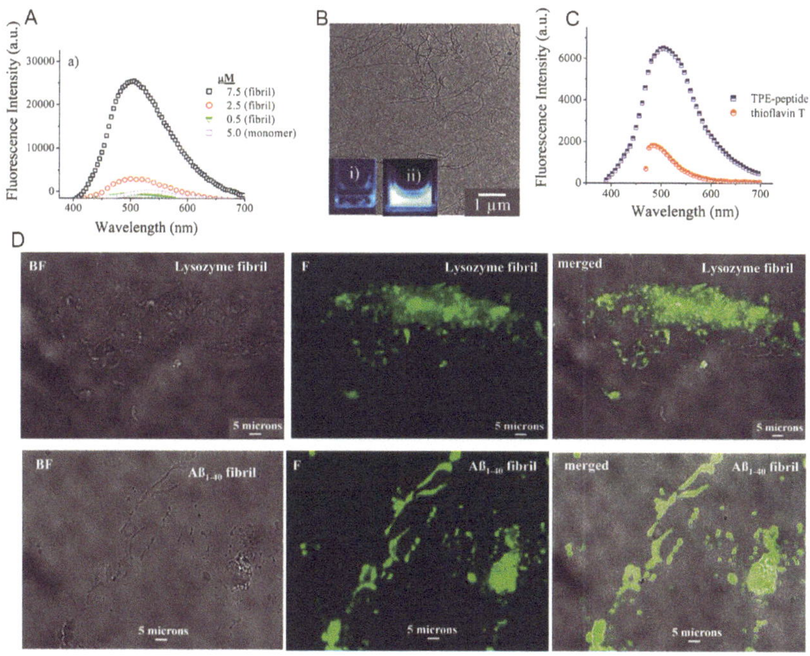

Fig. 9 (A) Fluorescence data of TPE-peptide in presence of Aβ-fibrils and monomers. (B) SEM images of insulin fibrils: (i) TPE-peptide solution in phosphate buffer at pH 7.4. (ii) Fluorescence light-up in presence of Aβ_{1-40} fibrils. (C) A comparison of efficacy of bioconjugate with ThT in the detection of Aβ-fibrils. (D) Images of TPE-peptide treated lysozyme fibrils and Aβ_{1-40} fibrils (BF-bright field and F-fluorescence). *Reproduced figures from Pradhan N, Jana D, Ghorai BK, Jana NR. Detection and monitoring of amyloid fibrillation using a fluorescence "switch-on" probe.* ACS Appl Mater Interfaces. *2015; 7:25813–25820. Copyright 2015 American Chemical Society.*

The authors compared the efficiency of bioconjugate with thioflavin-T in the detection of Aβ-fibrils. A superior performance was observed with the bioconjugate with better sensitivity, four times higher fluorescence than the conventional dye (Fig. 9C) and was unaffected by the presence of quencher ions and nanoparticles. The studies pointed out the advantages of the AIE active probe over the conventional dye ThT. A multifunctional biocompatible probe based on curcumin (Cur-N-BF_2; Fig. 4), showing AIE characteristics, was reported by Tang and co-workers.[42] The molecule exhibited greater fluorescence intensity in the solid-state compared to the solution-state. MTT assay on hippocampal neuronal cells (HT22 cells) gave low cytotoxicity compared to another analog of Cur-N-BF_2, curcumin, and ThT. An increased fluorescence was observed in the presence of Aβ_{1-42} fibrils formed in PBS buffer at 37 °C. ACQ effect was observed with ThT at 0.25 mM and curcumin at 0.05 mM. AIE probe gave intense fluorescence even at 0.4 mM concentration. Interestingly, a staining experiment carried out on fibrils from other proteins indicated the probe is selective only for Aβ_{1-42} fibrils. The efficacy and the selectivity of the probe in staining Aβ plaques in hippocampus and cerebral cortex regions of the brain slices (taken from transgenic (APP/PS1) mice) were proved by co-staining experiment with the commercial staining agent of β-amyloid antibody (Fig. 10A).

Based on fluorescence measurements, Cur-N-BF_2 was efficient in inhibiting Aβ fibrillation and was able to disassemble the fibrils. The results were further confirmed with TEM images and circular dichroism (CD) studies. A remarkable observation was that Cur-N-BF_2 could increase the cell viability of neuronal cells (HT22) which were pre-treated with Aβ fibrils (40 μM, 36 h). Thus, Cur-N-BF_2 possesses several remarkable features, ease in synthesis and high sensitivity over commercial dye, fluorescence-on detection of Aβ-amyloid deposits, inhibition of Aβ-fibrillation, disassembly of fibrils, and protection of neuronal cells (Fig. 10B). This theranostic multifunctional probe with high selectivity on amyloid-β fibrils is one of the promising candidates among diagnostics and therapeutics targeting amyloidosis.

Super-resolution imaging of amyloid-β plaques using AIE materials was reported by Tang group.[43] When compared to conventional optical microscopy techniques, super-resolution fluorescence microscopy provides better resolution images, allowing researchers to gain clear information. Four AIE-active molecules (DM-BZ, PD-BZ, PD-NA, and PD-NA-TEG; Fig. 4) with almost planar donor-acceptor structures were reported with high yields. They showed negligible fluorescence in THF solution but emitted strong

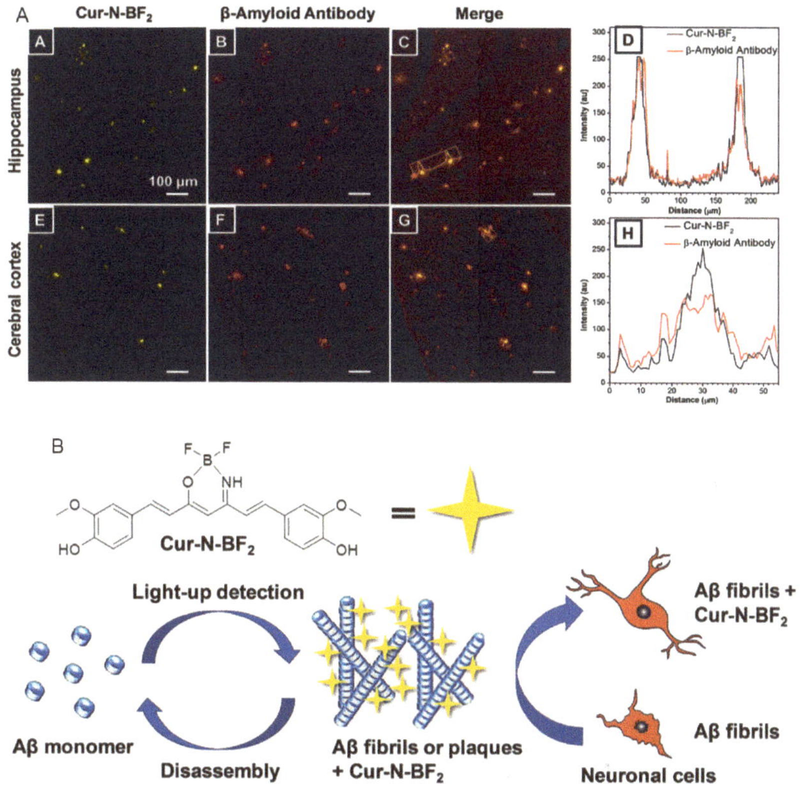

Fig. 10 (A) Confocal images of cerebral cortex and hippocampus regions (from APP/PS1 mice) stained with Cur-N-BF$_2$ (A and E) and β-amyloid antibody (B and F). (B) Multifunctional theranostic probe Cur-N-BF$_2$. *Reproduced from Yang Y, Li S, Zhang Q, et al. An AIE-active theranostic probe for light-up detection of Aβ aggregates and protection of neuronal cells.* J Mater Chem B. *2019;7:2434–2441, with permission from Royal Society of Chemistry.*

fluorescence from green to yellow (wavelength of 500–546 nm) in solid-state film and also in the aggregated state depicting their AIE characteristics.

They gave weak fluorescence in water containing 30 v/v% ethanol while after the addition of a small amount of fibrillar hen egg white lysosome (HEWL), fluorescence was switched on. The ex vivo affinity of these AIE probes for Aβ plaques were assessed using the brain tissue section from a transgenic (Tg) mouse (Tg6799, 9-month old, male) and with a reported fluorescent probe 8C as the dye for fluorescence colocalization (Fig. 11A and B). The colocalization images indicated that all the four compounds were excellent probes and among them, PD-NA and PD-NA-TEG gave better binding with Aβ plaques in the mouse brain. Compared to conventional fluorescence imaging, super-resolution imaging gave an extremely clear morphology of Aβ plaques. A lot of radiant nanofibers grow from the center indicates that amyloid deposit is initiated from the center (Fig. 11C). Such kind of probes can provide valuable information on the protein fibrillation pathway.

An ultrasensitive NIR probe with AIE characteristics in detecting amyloid-β deposits can be a promising candidate among the bioprobes for in vivo amyloid staining experiments.[44] With quinoline malononitrile group (QM), an NIR AIE-active probe QM-FN-SO_3 (Fig. 4) was reported recently by Zhu and co-workers. It has the following features (i) a lipophilic π-conjugated part to take care of the emission at the NIR region and blood-brain barrier (BBB) penetrability and (ii) a hydrophilic sulfonate group to prevent aggregation-induced emission before binding to amyloid-β deposits. The probe was non-emissive in an aqueous solution and became

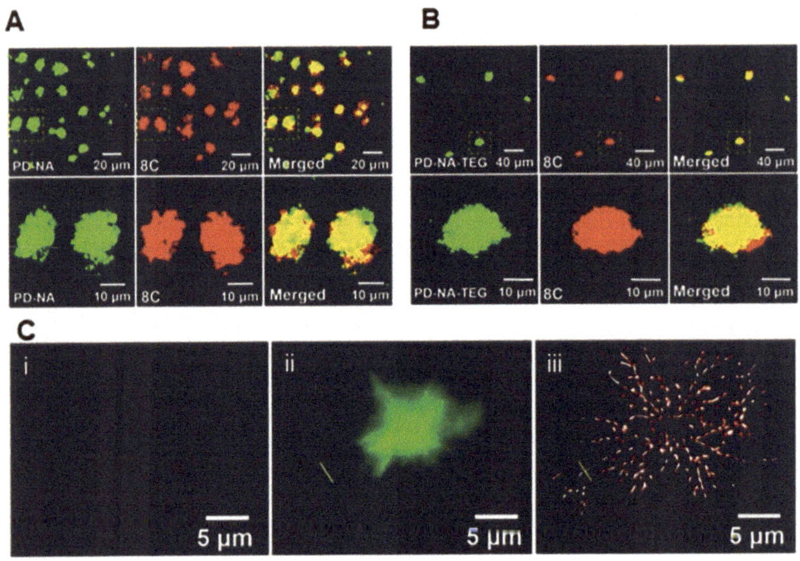

Fig. 11 Fluorescence colocalization imaging of Aβ deposits on brain slices from a transgenic (Tg) mouse (A) for PD-NA, (B) PD-NA-TEG. 8C is the reported fluorescent probe for colocalization. (C) Super-resolution imaging of Aβ deposits on brain slices from a Tg mouse with PD-NA-TEG. (i) bright-field image, (ii) conventional fluorescence imaging, and (iii) super-resolution imaging. *Reproduced from Wang YL, Fan C, Xin B, et al. AIE-based super-resolution imaging probes for β-amyloid plaques in mouse brain.* Mater Chem Front. *2018;2:1554–1562, with permission from Royal Society of Chemistry.*

emissive in an aggregated state with fluorescence at 720 nm, which demonstrated that it is a NIR probe with AIE characteristics. Further, it possessed superb stability at different pH and in fresh serum and exhibited an ultrahigh S/N ratio. For the sensitive detection of the amyloid deposits, binding affinity plays an important factor. In the displacement assay carried out by the addition of NIR probe to ThT/Aβ deposit complex, researchers observed a continuous decrease in fluorescence at 482 nm and an increase in fluorescence at 665 nm. This demonstrated the displacement of ThT by NIR probe to form QM-FN-SO_3/Aβ deposit complex. In the in vitro fluorescent staining experiment performed on Aβ plaques using QM-FN-SO_3 and thioflavin-S (ThS, a commonly used dye to stain amyloid-β plaques), staining was observed in slices of brain tissue from AD-model and no fluorescent labeling in the wild type mice brain. Also, the QM-FN-SO_3 exhibited a higher S/N ratio, better feedback showing that this NIR probe with AIE features can give "high fidelity Aβ plaque information." In the experiment performed to understand the BBB penetrability of QM-FN-SO_3, intravenous injection followed by HRMS analysis of brain homogenate indicated the presence of NIR-AIE probe, whereas the experiment with ThT did not give any presence of ThT in the brain homogenate. This demonstrated the BBB penetrability of the NIR-AIE probe. It was further supported by the in vivo imaging experiment after intravenous injection, a higher fluorescence intensity was observed in the brain region of AD-model mice than wild-type mice (Fig. 12).

An aggregation-induced emission enhancement (AIEE) bioprobe based on pyridothiazole moiety (PTC1; Fig. 4) was reported by Gour *et al.*[45] Aggregation properties were analyzed by dynamic light scattering measurements and AFM images. The fluorescence intensity increased with aggregation. Fluorescence quenching was observed with the addition of Cu^{2+} ions, and fluorescence reappeared in presence of amyloid fibrils. Further, the above result was confirmed with AFM study. AFM image of PTC1 alone showed the self-assembly with hairy-like morphologies. The self-assembly was disturbed with the addition of Cu^{2+} ions, and the assembly was regenerated with amyloid fibrils.

4.4 α-Synuclein

Oligomerization and fibrillation of α-synuclein (α-Syn) are associated with Parkinson's disease (PD), the second commonly observed neurodegenerative disease among the elderly population.[46,47] In vitro studies shows that

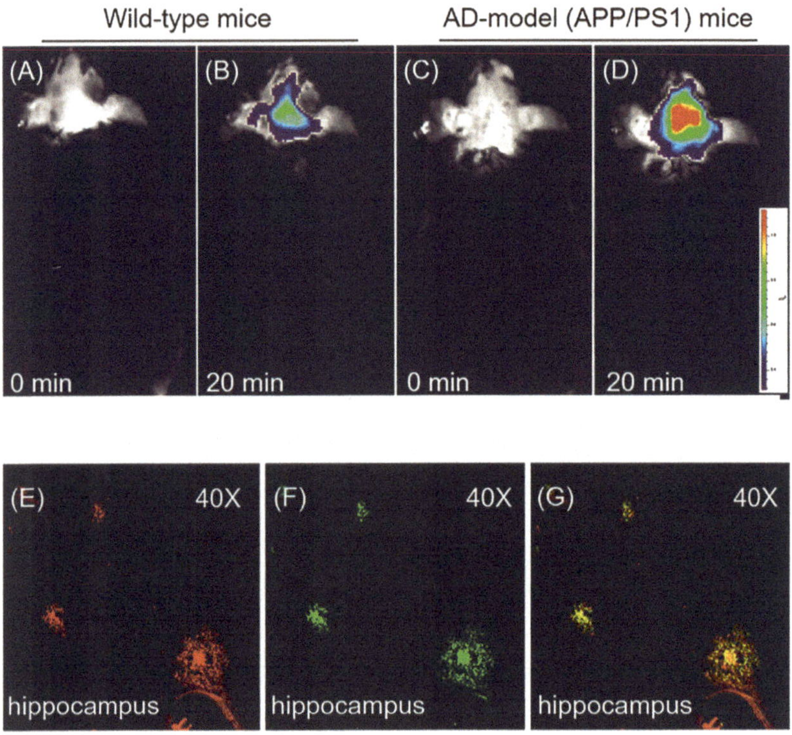

Fig. 12 In vivo fluorescence imaging after 20 min of intravenous injection of QM-FN-SO_3 (2 mg/kg) in wild-type mice (A and B) and AD model (APP/PS1) mice (C and D). Ex-vivo imaging of brain slice from AD model (APP/PS1) mice (E) after injection of the probe, (F) with antibody 254, and (G) merged. *Reproduced the figures with permission from Fu W, Yan C, Guo Z, et al. Rational design of near-infrared aggregation-induced-emission-active probes: in situ mapping of amyloid-β plaques with ultrasensitivity and high-fidelity.* J Am Chem Soc. *2019;141:3171–3177. Copyright 2019 American Chemical Society.*

fibril formation from α-Syn monomers follows a nucleated polymerization mechanism, as observed mainly with other amyloid-forming proteins.[48,49] The commonly used α-Syn fibril staining agent is ThT. One of the limitations of ThT is its inefficiency in detecting more cytotoxic early-stage conformers (oligomers). Tang and co-workers used an AIEgen comprising tetraphenylethene (TPE) attached to triphenyl phosphonium unit (TPE-TPP; Fig. 4), to follow the α-Syn fibrillation kinetics.[50] Compared to ThT, the probe exhibited remarkable fluorescence intensity in the presence of mature fibrils after 97 h (Fig. 13). A decrease in emission signal ($\lambda_{em}=$ 490 nm) of ThT bound to amyloid fibril upon addition of TPE-TPP was

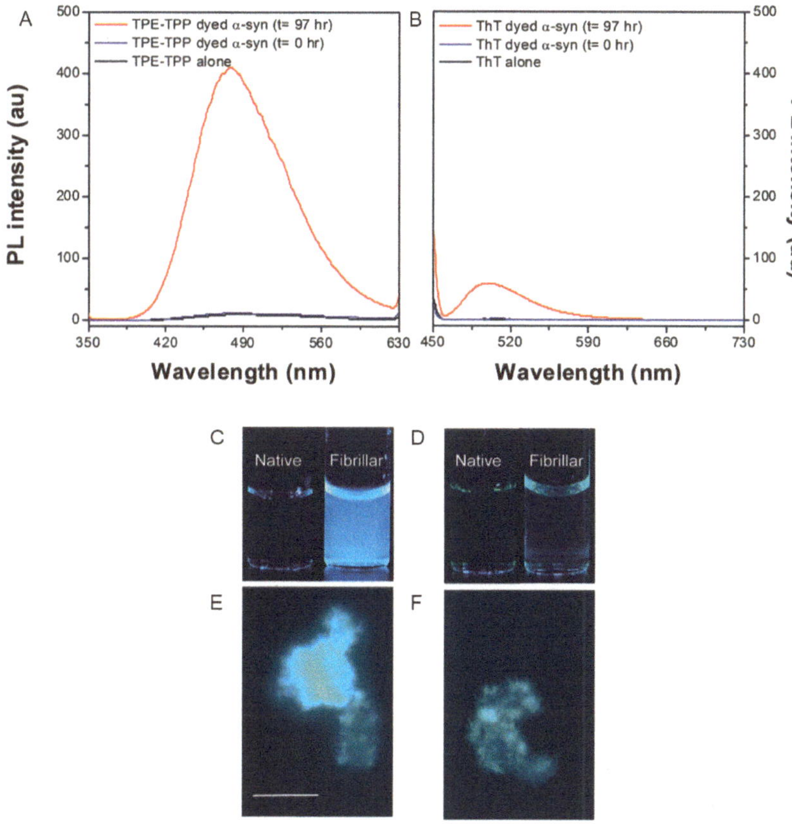

Fig. 13 (A) Emission spectra of TPE-TPP (15 μM, black) and with α-Syn (5 μM) at 0 h (blue) and after 97 h (red). (B) ThT and with fibrils under the same conditions. (C) Images were taken with UV light illumination ($\lambda = 365$ nm) after the addition of TPE-TPP to native and α-Syn fibrils. (E) Fluorescence staining of native form and fibrils with TPP-TPE and (F) with ThT. *Reproduced from Leung CWT, Guo F, Hong Y, et al. Detection of oligomers and fibrils of α-synuclein by AIEgen with strong fluorescence.* Chem Commun. *2015;51:1866–1869, with permission from Royal Chemical Society.*

observed in binding competition assay (ThT/TPE-TPP = 15 μM, α-Syn = 5 μM) indicated that TPE-TPP has a higher binding affinity than ThT toward fibrils. Fluorescence enhancement was observed at the early stage of fibrillation with TPE-TPP and not with ThT (at 22 h) indicated that AIEgen can detect intermediates (oligomers). This was confirmed by AFM, size exclusion chromatography, and dot blot. Thus, the use of TPE-TPP in monitoring kinetics of α-Syn fibrillation resulted in the identification of TPE-TPP as a promising fluorescent probe in monitoring amyloid fibrillation and detection of early-stage species (oligomers) and fibrils.

Recently, site-specific incorporation of AIE to amyloid-forming proteins (Aβ-peptide and α-synuclein; Fig. 4) was explored in the screening of amyloid inhibitors by Liu and co-workers.[51] An azide bearing unnatural amino acid (4-azido-L-phenylalanine, pAzF) was incorporated into the sequence of peptides. The click chemistry between the alkyne functional group on a water-soluble AIE molecule (2,2′-(((2-(4-ethynylphenyl)-2-phenylethene-1,1-diyl)bis(4,1-phenylene))bis(oxy))bis-(*N*,*N*,*N*-trimethylethanaminium) bromide, EPB) and azide functionality on the modified peptide resulted in AIE@amyloids. In the design strategy, phenylalanine and tyrosine residues were replaced with pAzF to minimize the conformation/aggregation perturbations. Using ThT as a fluorescence probe, the amyloidogenic property of the modified peptides was evaluated. Based on ThT assay and AFM images, three mutants were selected as they exhibited typical amyloidosis. The AIE moiety EPB was then introduced with cycloaddition reaction and the resultant EPB@amyloid probes exhibited fluorescence-on upon fibrillation. The aggregation-induced fluorescence observed from EPB@amyloid forming peptide upon fibrillation will be affected by the amyloid inhibitors. As a proof-of-concept, the probes were used as a platform to screen inhibitors. Tolcapone was found to have dual inhibitory activity on the amyloidosis with Aβ-peptide and α-synuclein. Probes with a suitable AIE at amyloid-forming proteins (without altering the conformation/aggregation tendency), can be useful in screening amyloid inhibitors.

4.5 Early-stage diagnostics

Recent studies in the amyloidosis process reveal that prefibrillar species are more cytotoxic than fibrils and play a key role in the pathological processes. The commonly used fluorescent molecules (ThT and Congo Red) fail to detect these species and hence in monitoring the early stages of fibrillation. Early diagnosis of fibrillation has high therapeutical relevance and few AIE fluorogens that can monitor the prefibrillar species have been reported.[52] As mentioned earlier, TPE-TPP, an AIEgen, is known to bind with fibrils of α-synuclein and in the detection of early-stage aggregates.[50] Compared to ThT, it shows more sensitivity and no self-quenching. In an in situ fluorescence assay with six different amyloid fibril forming proteins (RCM α-lactalbumin, hen egg-white lysozyme, RCM-κ-casein, α-synuclein, αB-crystallin, D76N β2-microglobulin), fluorescence enhancement was observed with the formation of prefibrillar species, indicating that dye could effectively monitor the prefibrillar species. It was further confirmed by the

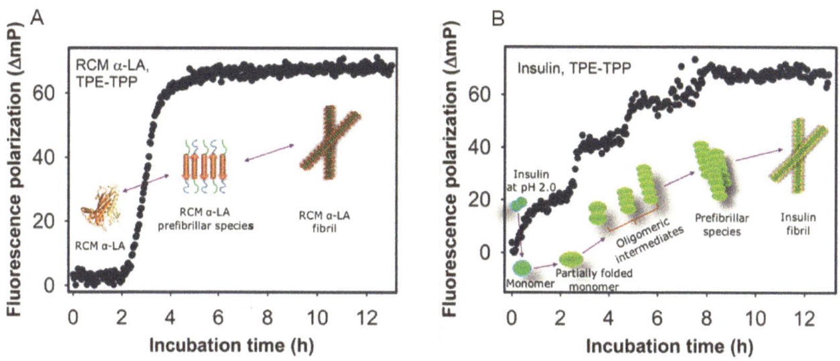

Fig. 14 FP assay with TPP-TPE in fibrillation of (A) RCM-αLA (125 μM), (B) insulin (200 μM). *Reproduced figures from Kumar M, Hong Y, Thorn DC, Ecroyd H, Carver JA. Monitoring early-stage protein aggregation by an aggregation-induced emission fluorogen.* Anal Chem. *2017;89:9322–9329. Copyright 2017 American Chemical Society.*

in situ fluorescence polarization (FP) experiment (Fig. 14) Fluorescence enhancement was observed only when the protein underwent fibrillation and not with the amorphous aggregates and molten globule state. Thus, TPP-TPE proves to be a superior diagnostic probe for the detection of prefibrillar assemblies and in monitoring amyloidosis.

5. Perspectives

In the last 10 years, substantial progress is observed in the development of AIE probes in bioimaging. AIE probes for monitoring amyloid kinetics and detection of amyloid aggregates are one among them. These promising molecules can provide new insights into the complicated amyloid process and we can envisage that they will be able to create a remarkable impact in translational research. AIE probes in detecting different amyloid fibrils, monitoring amyloid kinetics, and early-stage conformers are known. Other notable features are they can be modified as NIR probes, multifunctional probes, and can be used in super-resolution imaging. In few cases, they exhibit both diagnostic and inhibitory properties. Despite the progress attained with diverse fluorescent probes, challenges and difficulties still exist in tackling amyloidosis diseases. Research needs to be extended to other important amyloid-forming proteins such as tau protein. Diagnostic probes and inhibitors for tau protein aggregates are important in AD therapy. Most of the amyloid staining probes described here function in vitro. Recent analysis shows that the structure of in vivo formed AD fibrils was different from the in vitro formed fibrils. We can anticipate that in the coming years, a greater number of efficient probes that can work in vivo conditions and in early-stage diagnosis will be available. Efforts may be concentrated on the development of NIR probes with good selectivity, sensitivity, and BBB penetrability. This will not only help in unveiling the mystery associated with amyloidosis but also help in the development of potent drugs.

Acknowledgments

The authors acknowledge the UGC-BSR grant (394/2017/BSR) and the SERB grant (ECR/2018/002434).

Conflicts of interest

There are no conflicts to declare.

References

1. Dobson CM. Protein folding and misfolding. *Nature*. 2003;426:884–890.
2. Dobson CM. The amyloid phenomenon and its links with human disease. *Cold Spring Harb Perspect Biol*. 2017;9:1–14.
3. Almeida ZL, Brito RMM. Structure and aggregation mechanisms in amyloids. *Molecules*. 2020;2:1195.
4. Chiti F, Dobson C. Protein misfolding, amyloid formation, and human disease: a summary of progress over the last decade. *Annu Rev Biochem*. 2017;86:27–68.
5. Pandey G, Morla S, Kumar S, Ramakrishnan V. Modulation of tau protein aggregation using 'Trojan' sequences. *Biochim Biophys Acta—Gen Subj*. 2020;1864, 129569.
6. Aliyan A, Cook NP, Martí AA. Interrogating amyloid aggregates using fluorescent probes. *Chem Rev*. 2019;119:11819–11856.
7. Lv G, Shen Y, Zheng W, Yang J, Li C, Lin J. Fluorescence detection and dissociation of amyloid-β species for the treatment of Alzheimer's disease. *Adv Ther*. 2019;2:1900054.
8. Guo M, Song H, Li K, et al. A new approach to developing diagnostics and therapeutics: aggregation-induced emission-based fluorescence turn-on. *Med Res Rev*. 2020;40: 27–53.
9. Gyasi YI, Pang YP, Li XR, et al. Biological applications of near infrared fluorescence dye probes in monitoring Alzheimer's disease. *Eur J Med Chem*. 2020;187, 111982.
10. Eisenberg DS, Sawaya MR. Structural studies of amyloid proteins at the molecular level. *Annu Rev Biochem*. 2017;86:69–95.
11. Fitzpatrick AWP, Debelouchina GT, Bayro MJ, et al. Atomic structure and hierarchical assembly of a cross-β amyloid fibril. *Proc Natl Acad Sci USA*. 2013;110:5468–5473.
12. Li D, Liu C. Structural diversity of amyloid fibrils and advances in their structure deter mination. *Biochemistry*. 2020;59:639–646.
13. Reddy JP, Swain D, Pedireddi VR. Polymorphism and phase transfer behavior of solid forms of 4-amino-3,5-dintrobenzamide. *Cryst Growth Des*. 2014;14:5064–5071.
14. Reddy JP, Prabhakaran P, Steed JW. Polymorphism of *Z*-(3-bromopropenoic acid): a high and low *Z*′ pair. *Cryst Growth Des*. 2016;16:4021–4025.
15. Tompa P, Fuxreiter M. Fuzzy complexes: polymorphism and structural disorder in protein-protein interactions. *Trends Biochem Sci*. 2008;33:2–8.

16. Kollmer M, Close W, Funk L, et al. Cryo-EM structure and polymorphism of Aβ amyloid fibrils purified from Alzheimer's brain tissue. *Nat Commun*. 2019; 10:4760.
17. Toyama BH, Weissman JS. Amyloid structure: conformational diversity and consequences. *Annu Rev Biochem*. 2011;80:557–585.
18. Arosio P, Knowles TPJ, Linse S. On the lag phase in amyloid fibril formation. *Phys Chem Chem Phys*. 2015;17:7606–7618.
19. Stefani M. Structural features and cytotoxicity of amyloid oligomers: implications in Alzheimer's disease and other diseases with amyloid deposits. *Prog Neurobiol*. 2012; 99:226–245.
20. Teoh CL, Su D, Sahu S, et al. Chemical fluorescent probe for detection of Aβ oligomers. *J Am Chem Soc*. 2015;137:13503–13509.
21. Zhu H, Hamachi I. Fluorescence imaging of drug target proteins using chemical probes. *J Pharm Anal*. 2020;10:426–433.
22. Gu X, Kwok RTK, Lam JWY, Tang BZ. AIEgens for biological process monitoring and disease theranostics. *Biomaterials*. 2017;146:115–135.
23. Martinić I, Eliseeva SV, Petoud S. Near-infrared emitting probes for biological imaging: organic fluorophores, quantum dots, fluorescent proteins, lanthanide(III) complexes and nanomaterials. *J Lumin*. 2017;189:19–43.
24. Wang YF, Zhang T, Liang XJ. Aggregation-induced emission: lighting up cells, revealing life! *Small*. 2016;12:6451–6477.
25. Hong Y, Lam JWY, Tang BZ. Aggregation-induced emission. *Chem Soc Rev*. 2011;40: 5361–5388.
26. Eisert R, Felau L, Brown LR. Methods for enhancing the accuracy and reproducibility of Congo red and thioflavin T assays. *Anal Biochem*. 2006;353:144–146.
27. Mei J, Leung NLC, Kwok RTK, Lam JWY, Tang BZ. Aggregation-induced emission: together we shine, united we soar! *Chem Rev*. 2015;115:11718–11940.
28. Ding D, Li K, Liu B, Tang BZ. Bioprobes based on AIE fluorogens. *Acc Chem Res*. 2013;46:2441–2453.
29. Niu G, Zhang R, Shi X, et al. AIE luminogens as fluorescent bioprobes. *TrAC - Trends Anal Chem*. 2020;123, 115769.
30. Nilsson MR. Insulin amyloid at injection sites of patients with diabetes. *Amyloid*. 2016;23:139–147.
31. Nielsen L, Frokjaer S, Brange J, Uversky VN, Fink AL. Probing the mechanism of insulin fibril formation with insulin mutants. *Biochemistry*. 2001;40:8397–8409.
32. Hong Y, Meng L, Chen S, et al. Monitoring and inhibition of insulin fibrillation by a small organic fluorogen with aggregation-induced emission characteristics. *J Am Chem Soc*. 2012;134:1680–1689.
33. Huang Q, Xie J, Liu Y, Zhou A, Li J. Detecting the formation and transformation of oligomers during insulin fibrillation by a dendrimer conjugated with aggregation-induced emission molecule. *Bioconjug Chem*. 2017;28:944–956.
34. Uddin A, Roy B, Jose GP, Hossain SS, Hazra P. Sensing and modulation of amyloid fibrils by photo-switchable organic dots. *Nanoscale*. 2020;12:16805–16818.
35. Tong H, Lou K, Wang W. Near-infrared fluorescent probes for imaging of amyloid plaques in Alzheimer's disease. *Acta Pharm Sin B*. 2015;5:25–33.
36. Ding S, Yao B, Schobben L, Hong Y. Barbituric acid based fluorogens: synthesis, aggregation-induced emission, and protein fibril detection. *Molecules*. 2020;25:32.
37. Hamley IW. The amyloid beta-peptide: a chemist's perspective. role in Alzheimer's and fibrillization. *Chem Rev*. 2012;112:5147–5192.
38. Palmqvist S, Mattsson N, Hansson O. Cerebrospinal fluid analysis detects cerebral amyloid-β accumulation earlier than positron emission tomography. *Brain*. 2016;139: 1226–1236.

39. Wang Y, Qiu Y, Sun A, et al. Dual-functional AIE fluorescent probes for imaging β-amyloid plaques and lipid droplets. *Anal Chim Acta*. 2020;1133:109–118.
40. Farmer BC, Walsh AE, Kluemper JC, Johnson LA. Lipid droplets in neurodegenerative disorders. *Front Neurosci*. 2020;14:742.
41. Pradhan N, Jana D, Ghorai BK, Jana NR. Detection and monitoring of amyloid fibrillation using a fluorescence "switch-on" probe. *ACS Appl Mater Interfaces*. 2015;7: 25813–25820.
42. Yang Y, Li S, Zhang Q, et al. An AIE-active theranostic probe for light-up detection of Aβ aggregates and protection of neuronal cells. *J Mater Chem B*. 2019;7:2434–2441.
43. Wang YL, Fan C, Xin B, et al. AIE-based super-resolution imaging probes for β-amyloid plaques in mouse brain. *Mater Chem Front*. 2018;2:1554–1562.
44. Fu W, Yan C, Guo Z, et al. Rational design of near-infrared aggregation-induced-emission-active probes: in situ mapping of amyloid-β plaques with ultrasensitivity and high-fidelity. *J Am Chem Soc*. 2019;141:3171–3177.
45. Gour N, Kshtriya V, Gupta S, et al. Synthesis and aggregation studies of a pyridothiazole-based AIEE probe and its application in sensing amyloid fibrillation. *ACS Appl Bio Mater*. 2019;2:4442–4455.
46. Goedert M, Jakes R, Spillantini MG. The synucleinopathies: twenty years on. *J Parkinsons Dis*. 2017;7:S51–S69.
47. Stefanis L. α-Synuclein in Parkinson's disease. *Cold Spring Harb Perspect Med*. 2012; 2, a009339.
48. Conway KA, Lee SJ, Rochet JC, Ding TT, Williamson RE, Lansbury PT. Acceleration of oligomerization, not fibrillization, is a shared property of both α-synuclein mutations linked to early-onset Parkinson's disease: implications for pathogenesis and therapy. *Proc Natl Acad Sci USA*. 2000;97:571–576.
49. Frey B, AlOkda A, Jackson MP, Riguet N, Duce JA, Lashuel HA. Monitoring alpha-synuclein oligomerization and aggregation using bimolecular fluorescence complementation assays: what you see is not always what you get. *J Neurochem*. 2021; 157:872–888.
50. Leung CWT, Guo F, Hong Y, et al. Detection of oligomers and fibrils of α-synuclein by AIEgen with strong fluorescence. *Chem Commun*. 2015;51:1866–1869.
51. Jia L, Wang W, Yan Y, et al. General aggregation-induced emission probes for amyloid inhibitors with dual inhibition capacity against amyloid β-protein and α-synuclein. *ACS Appl Mater Interfaces*. 2020;12:31182–31194.
52. Kumar M, Hong Y, Thorn DC, Ecroyd H, Carver JA. Monitoring early-stage protein aggregation by an aggregation-induced emission fluorogen. *Anal Chem*. 2017;89: 9322–9329.

CHAPTER SIX

AIE active polymers for biological applications

Nehal Zehra[a], Arvin Sain Tanwar[a], Mst Nasima Khatun[a], Laxmi Raman Adil[a], and Parameswar Krishnan Iyer[a,b,c,*]
[a]Department of Chemistry, Indian Institute of Technology Guwahati, Guwahati, Assam, India
[b]Centre for Nanotechnology, Indian Institute of Technology Guwahati, Guwahati, Assam, India
[c]School of Health Sciences and Technology, Indian Institute of Technology Guwahati, Guwahati, Assam, India
*Corresponding author: e-mail address: pki@iitg.ac.in

Contents

Abstract

The discovery of aggregation-induced emission (AIE) phenomenon, significantly altered the understanding of the scientific world about the luminophore aggregation. Polymers with AIE features have recently emerged as promising materials with wide range of applications in optoelectronics devices, chemosensors, bioimaging, cancer theranostics and drug delivery. By introducing the AIE active molecule into the polymer structure, novel materials encompassing the characteristics properties of both the functional materials such as excellent brightness, versatile structure modification, high biocompatibility, exceptional stability and facile processability are achieved. This chapter presents the advances in synthetic design as well as potential biological applications of AIE active polymers, beginning with a brief introduction to the AIE phenomenon. The versatile

Progress in Molecular Biology and Translational Science, Volume 185
ISSN 1877-1173
https://doi.org/10.1016/bs.pmbts.2021.07.002

synthetic route, easier functionalization, and light up feature of the AIE active polymers offer direct visualization of the physiological processes within or outside the living organisms. This chapter also precisely describes the photodynamic therapy/photothermal therapy (PDT/PTT) with up-to-date advancement of AIE active polymer and their emerging applications in biomedical field. The AIE active Photosensitizers (PSs) are much more efficient in singlet oxygen ($^1O_2$) production than their small molecule AIE active PSs due to their enhanced inter system crossing (ISC) process and improved light-harvesting ability. Additionally, the present chapter aims to focus on all recent AIE active polymers for drug screening and drug delivery. The AIE active polymer often shows decent drug loading capacity, high stability and good biocompatibility comprising image guided drug monitoring features. Lastly, the concluding discussion reveals the future prospective of the AIE active polymers.

1. Introduction

1.1 General introduction of AIE active polymer

Organic luminophores have undoubtedly emerged as indispensable tools for real-time monitoring of biological analytes and dynamic processes in the living systems. The development of new luminophores has not only advanced the area of bioimaging but also revolutionized medical diagnostic and therapeutics. The luminophores are designed and explored as bioprobes, photosensitizers, antimicrobial, anticancer agents, and images guided therapeutics based on their characteristics optical properties, and spatiotemporal resolution. However, the potential use of conventional luminophores are narrowed in terms of sensitivity, effectiveness, and stability due to the widely experiential phenomenon; aggregation caused quenching (ACQ). In ACQ effect, the luminophores exhibit strong emission in dilute solution while the emission gets quenched in the aggregate or solid-state. The mechanism of such concentration caused quenching was first explained by Birks in his book entitled "*Photophysics of Aromatic Molecules*." Generally, the organic luminophores have planar structures and experience strong intermolecular π-π interaction in their condensed states. The aromatic core of adjacent planar luminophores can interlock themselves in H-type of aggregates favoring the formation of excimer or exciplexes with non-radiative relaxation. The aggregation quenching effect limits the application scope of organic luminophores as the sensory materials and molecular imaging probes at their higher concentrations. For instance, the sensitivity of biosensor is determined by fluorescence change after analyte binding.[1] However, the brightness or fluorescence of luminophore in a very dilute solution is often less, ensuring the biosensor's reduced sensitivity. Furthermore, the dilute solutions of the bioprobes are more susceptible to photobleaching when exposed to the harsh excitation laser. The ACQ plays a detrimental effect on labeling the luminophore to biomolecules and compels them to work in a dilute solution.[2–5]

In 2001, Tang et al. discovered a photophysical phenomenon of fluorophore aggregation, which now, is popularly known as aggregation-induced emission (AIE). In the AIE effect, the fluorophore is almost non-emissive in the isolated molecular state, but exhibits boosted emission in the condensed state or aggregate state. The molecule or luminophore having AIE property is termed as AIEgen. The AIEgens have twisted shaped structures with free rotating or vibrating structural component. In a molecularly dissolved state, the exciton energy is consumed and dissipated non-radiatively by the dynamic intramolecular motion of rotatable or vibratable groups. However, in the aggregate form or condensed state, such free motions are restricted due to the physical hindrance of neighboring molecules, facilitating radiative relaxation of the exciton. The AIE effect has vast practical implications for developing "light up" biosensors and bioassays that work at any concentrations. AIE biosensors have higher sensitivity, superior photostability, and signal reliability than traditional ACQ counterparts. The AIEgens are also excellent exogenous contrast agents for labeling biomolecules and tracking dynamics processes in the living system. Furthermore, the AIEgens have high brightness and low background signal, which endows them with wash free imaging of cell organelles, microbes, 3D multicellular spheroids.[6–8]

The AIE active polymers are an important emerging research discipline with many possible opportunities. They are likely to be more beneficial compared to small molecular AIEgens. For instance, the polymer's molecular structure, functionalities, morphology, and topology can easily be fine-tuned compared to small-molecule systems. They also exhibits boosted emission as the polymer's sides-chains or backbone imposes steric hindrance on the incorporated AIE segment. Thus, incorporating AIE unit into the polymer structure is beneficial for the development of novel functional materials with unique optoelectronic properties, higher mechanical strength, and good processability. For instance, the AIE active conjugated polymers offer ultra-sensitive response as they exhibit an efficient exciton transfer pathway over the entire conjugated backbone, resulting in signal amplification from several conjugated units. Similarly, the AIE active non-conjugated polymers exhibits amphiphilicity, biocompatibility and stimuli response properties and thus, widely explored for *in vivo/in vitro* bioimaging, drug delivery and image guided therapy.[9,10] This chapter presents the importance of AIE active polymers in biological science, with focus on biological imaging, sensing, and biomedical applications.

1.2 Synthesis and properties of AIE active polymers

The AIE active polymer can be strategically designed through various synthetic methods. The basic principle is to insert, link, or graft representative AIEgens into the polymer main or side chain. They can be constructed in the following ways:

(1) AIE active non-conjugated polymer with AIEgens at side chain can be generated by (a) homopolymerization or (b) copolymerization of the monomer containing AIEgens as pendant group.

(2) AIE active conjugated polymer with AIEgens at polymeric backbone can be constructed through (a) homopolymerization of AIE active monomer or, (b) copolymerization of AIE active monomer with other monomer.

Generally, the chain polymerization, i.e., vinyl homo and copolymerization has been widely used to construct AIE active non-conjugated polymers. Step polymerization such as polycondensations, and poly coupling reactions (Stille, Suzuki, and Sonogashira) have been utilized to furnish AIE active conjugated polymers. This section will present the various synthetic route of AIE active polymers as well as summarizes their photophysical properties (Table 1).

1.2.1 Chain or addition polymerization

The well-known free radical chain polymerization is commonly employed for AIE active vinyl monomer to form an AIE active homopolymer or copolymer. The C—C double bond is highly reactive and, thus, susceptible to the free radical attack. Once chain radical is generated, more monomer units are added to the single propagating chain resulting in high molecular weight polymers. The resulting polymer displays excellent solubility and weak emission in the organic solvent (dissolved state) while its emission gets boosted in water (aggregate state). The polymer skeleton is hydrophobic in nature and thus aggregates in water. The aggregation of the polymer greatly restricts intramolecular motions forcing the exited state luminogens to decay radiatively. These AIE active polymers have been utilized for sensing and imaging application in virtue of their excellent photophysical properties and biocompatibility. A cationic AIE copolymer was synthesized by two-step radical chain polymerization using tetraphenylene ethylene linked vinyl

Table 1 Characteristics properties of AIE active polymers.

Polymer	Mol. Wt. Da	Excitation (λ_{ex}) agg, nm	Emission (λ_{em})agg, nm	Lifetime (ns)	Quantum yield (ϕ)	Size (nm)	Application	References
P1	22,600	334	480	1.24, 5.70	0.12	>100	Detection of glucose	11
P2	na	438	580	na	na	192	Bioimaging of human lung carcinoma cells	12
P3	19,000	330	470	na	na	141.2	Bioimaging and cancer therapy	13
P4	13,800	340 nm	479 nm	na	na	130	*In vivo* drug release monitoring	14
P5–P7	12,000, 8300, 14,000	404,401,404	553,557,564	na	0.03,0.09,0.07	na	H_2O_2 chemosensing	15
P8	8600	356	521	na	0.562	na	Selective viable cell discrimination	16
P9	131,700	800	730	na	na	45	Two photon bioimaging	17
P10–P18	2800–6300	450–740	685–1090	na	0.019–0.054	14–24	NIR-IIa fluorescence for mouse-brain imaging	18
P19–P21	6990, 8670	370	567,600,517	1.08, 1.09	na	50	Multicolor drug delivery system	19
P22–P23	10,800, 15,500	400	660	na	0.012,0.003	29, 30	Two photon photodynamic therapy	20

Continued

Table 1 Characteristics properties of AIE active polymers.—cont'd

Polymer	Mol. Wt. Da	Excitation (λ_{ex}) agg, nm	Emission (λ_{em})agg, nm	Lifetime (ns)	Quantum yield (ϕ)	Size (nm)	Application	References
P24	na	429	621	na	0.016	103	Photodynamic therapy and antibacterial application	21
P25–P27	2000,4100,11,000	458,490,440	616,660,620	na	0.07	~100	Photodynamic therapy	22
P28	2300	500	700	4.9	0.09	26	Imaging of H_2O_2 inflammation	23
P29–P32	2700, 5900, 9200, 4800	870,700,663,709	1112,1010,907,1080	na	0.03,0.01,0.02	100	Near NIR fluorescence imaging	24
P33	144,000	415	535	na	0.2	na	DNA hybridization detection	25
P34	na	408	530	na	0.14	20	Mitochondria targeting turn on sensing and imaging of Cu^{2+} and H_2S	26
P35	1,160,000	338	479	na	na	na	Long-term cellular tracking	27
P36	na	350	490	na	na	100	Monitoring of drug release and cancer therapy	28
P37	na	488	672	na	0.1	na	Multimodal theranostics application	29

monomer, azobisisobutyronitrile (AIBN) as initiator. Through the first radical polymerization, an oligomer of styrene and TPE unit was obtained and further polymerized with vinyl pyridine to get the polymer **P1** (Scheme 1).

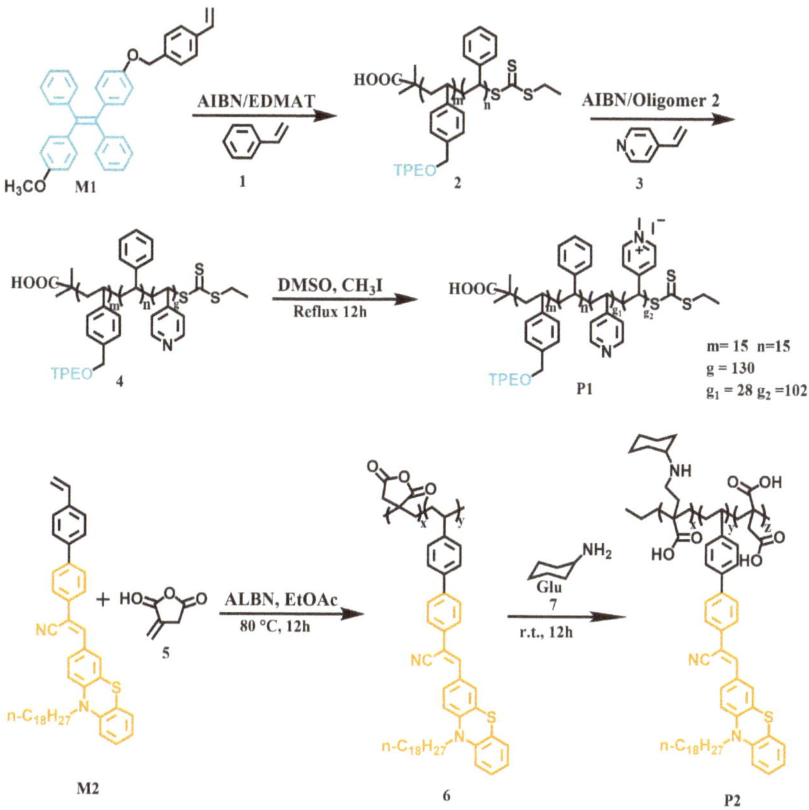

Scheme 1 Synthetic route of AIE active polymers **P1–P2** through free radical polymerization.

The polymer showed weak emission in DMF and DMSO: H_2O fraction. This may be ineffective self-assembly of the **P1** in DMSO: H_2O fraction resulting in the TPE moiety's unrestricted molecular motion. However, polymer shows boosted emission in its micellar form with four times higher fluorescence intensity than the DMSO: H_2O solution.[11] Similarly, another AIE active glycopolymer **P2** was reported for imaging of human lung carcinoma cells. The polymer **P2** was synthesized by free radical chain polymerization between AIE active phenothiazine monomer and biobased itaconic anhydride followed by ring-opening reaction of itaconic anhydride with glucosamine hydrochloride (Scheme 2). The **P2** self-assembles into

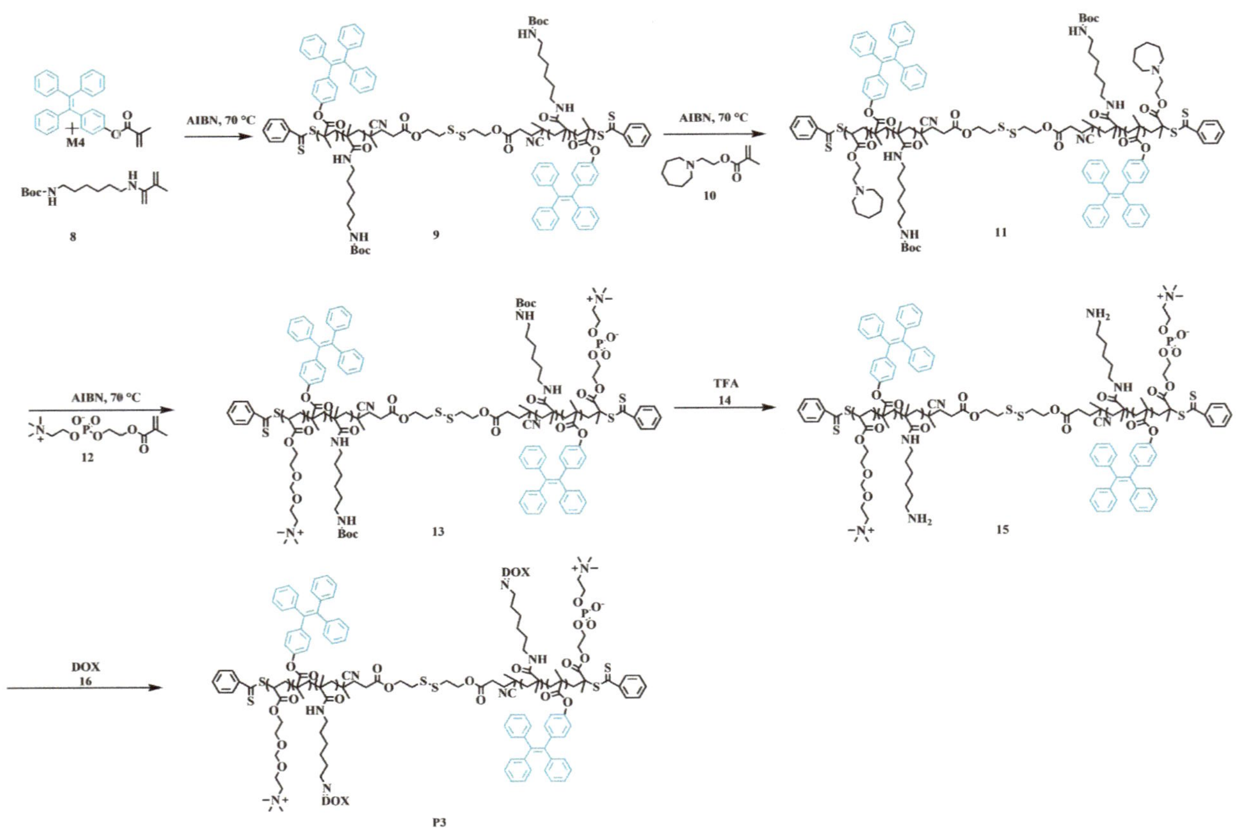

Scheme 2 Synthesis of AIE active amphiphilic polymer **P3** through free radical polymerization.

highly emissive nanoparticle (NP) in an aqueous solution with hydrophilic glycosyl and carboxyl groups at the surface and hydrophobic aromatic group at the core.[12]

Doxorubicin conjugated amphiphilic AIE active polymer **P3** for bio-imaging and cancer therapy has also been prepared via step by step reversible addition/fragmentation chain transfer (RAFT) polymerization (Scheme 2). The polymer **P3** consists of pH-responsive poly(2-azepane ethyl methacrylate) (PAEMA), zwitterionic poly(2-methacryloyloxyethyl phosphorylcholine) (PMPC), doxorubicin drug, and TPE as a luminophore. The polymer **P3** forms highly fluorescent core-shell micelle in aqueous solution with TPE and doxorubicin as core and hydrophilic PAEMA as a shell.[13]

1.2.2 Step or condensation polymerization

A condensation polymerization is a type of step-growth polymerization in which two different bifunctional or polyfunctional monomers react to form a large molecular weight product with the elimination of a small molecule as a byproduct. The condensation polymers can be synthesized through various synthetic routes such as amidation, esterification, polycoupling reactions (Stille, Suzuki, Knoevenagel). The condensation polymerizations are exploited to develop AIE active polyamides, polyurethanes with outstanding physical and photophysical properties. For example, an AIE active thermogelling polyurethanes **P4** was reported as a self-indicting drug delivery system for real-time and long term tracking of drug release in the tumor (Scheme 3). The polyurethane **P4** was synthesized from poly(ethylene glycol)-PEG, poly(propylene glycol)-PPG, hexamethylene diisocyanate, and TPE diol derivative. The **P4** forms weakly emissive self-assembled micellar structure at Critical Micellar Concentration (CMC) and highly fluorescent gel above CMC. This anomaly results from fortifying the TPE unit's rigidity in the close-packed 3D network structure of the gel. The **P4** gel is sensitive to the temperature and undergoes the transition from micelle to gel state at a higher temperature.[14] Recently, condensation polymers have also been constructed using poly coupling reactions such as Suzuki, Stille, Sonogashira, Knoevenagel, etc. Through the transition metal-catalyzed polycoupling reactions, AIE active conjugated polymers can be designed. Some recent examples of AIE active polymers prepared through polycoupling reactions are discussed.

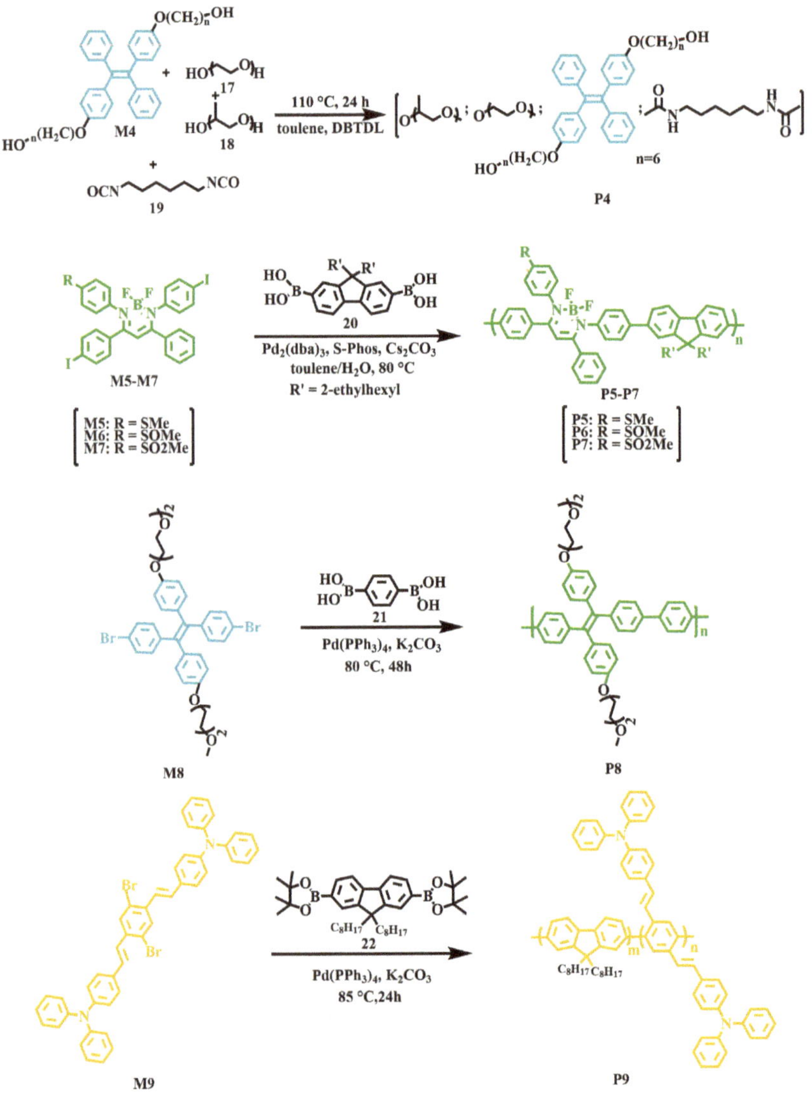

Scheme 3 Synthetic route of AIE active polymers **P4–P9** via condensation polymerization.

1.2.2.1 Suzuki reaction

Polycoupling reactions between aromatic diboronic acid or ester and dibromides are commonly explored approaches to construct AIE active conjugated polymers. As shown in Scheme 2, a series of organoborane-based AIE active conjugated polymers **P5–P7** were designed using the Suzuki cross-coupling reaction (Scheme 3). The sulfide or sulfoxide substituted boron diiminates was used as AIE monomer and coupled with the ACQ, fluorene diboronic acid. The photophysical properties of **P5–P7** AIE active conjugated polymers were fine-tuned by the insertion of sulfide or sulfoxide group at boron diiminates monomer. The sulfoxides containing boron diminates polymer **P6** and **P7** displays red-shifted charge transfer emission compared to the **P5** polymer. This is due to the electron-withdrawing nature of the sulfoxides. The polymers show AIE behavior in the aggregate state due to the restriction of phenyl rings' intramolecular motion in boron diminates monomer.[15] Similarly, a copolymer of TPE and phenylene **P8** was constructed for the differentiation of viable cells (Scheme 3).[16] In another study, a NIR emitting copolymer of triphenylamine derivative and fluorene **P9** was prepared via the Suzuki poly coupling reaction for the two-photon bioimaging application (Scheme 3). The hybrid nanoparticle of **P9** with perylene diimide derivative (PDI) was prepared for two-photon excited NIR emission via energy transfer between the **P9** and PDI derivative.[17] In a recent report, a series of AIE active polymers of phenothiazine **P10–P18** were prepared via the Suzuki coupling reaction. The phenothiazine unit has nonplanar structure with a dihedral angle of 153° due to sp^3 N and O atoms. Moreover, it is an electron donor and can be coupled with the strong acceptor to get the NIR emissive narrow bandgap polymers as shown in (Scheme 4).[18] In a similar fashion, multicolor AIE active polymers **P19–P21** using a molecular engineering approach have also been reported (Scheme 4). The **P19–P21** has the same acceptor unit (BODIPY) in its main chain, while the donor unit differs. The **P19** and **P20** exhibit AIE character; however, **P21** does not show significant enhancement in the fluorescence in the aggregate state. The ACQ nature of **P21** is due to the strong ICT state. The **P19** and **P20** were adapted for self-illuminating tumor-targeting drug delivery systems.[19] The polymerization between TPE diboronic ester and 2-(bis(4-bromophenyl)methylene)malononitrile affords AIE active conjugated polymers **P22** and **P23** (Scheme 4). Both the polymers **P22** and **P23** are fabricated into nanoparticles and showed bright NIR emission, high two-photon absorption, and singlet oxygen generation efficiency.[20] Later on, AIE TPE-based copolymer were also designed via Suzuki reaction, starting from dibromo substituted TPE derivative as electron donor, benzothiadiazole as an electron acceptor. The polymer **P24** was post functionalized with the 4-azidoperfluorobenzoate as a receptor for selectively binding with the

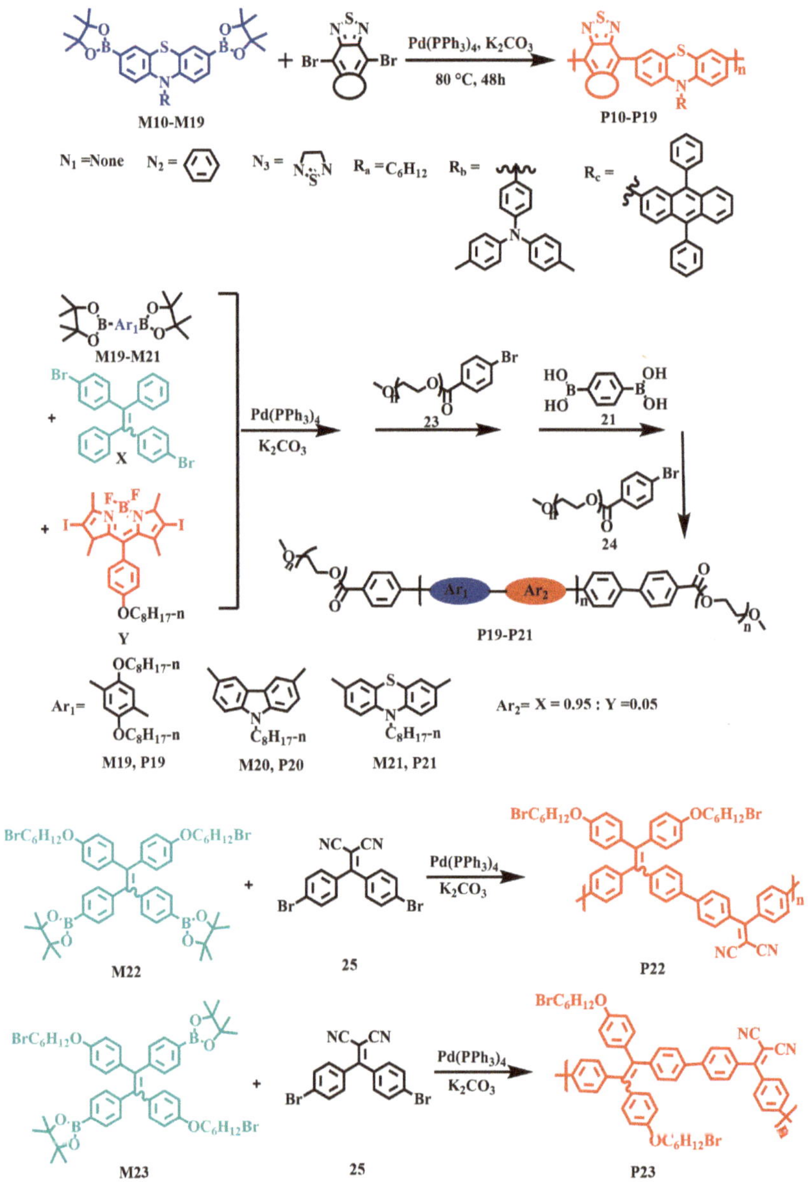

Scheme 4 Synthetic route of AIE active polymers **P10–P23** via condensation polymerization.

bacteria cell (Scheme 5). The **P24** showed boosted NIR emission and singlet oxygen generation in aggregate state upon illumination of light. The **P24** was utilized for the selective killing of bacteria cells over the mammalian cell.[21] Similarly, a series of AIE active donor-acceptor type of copolymers

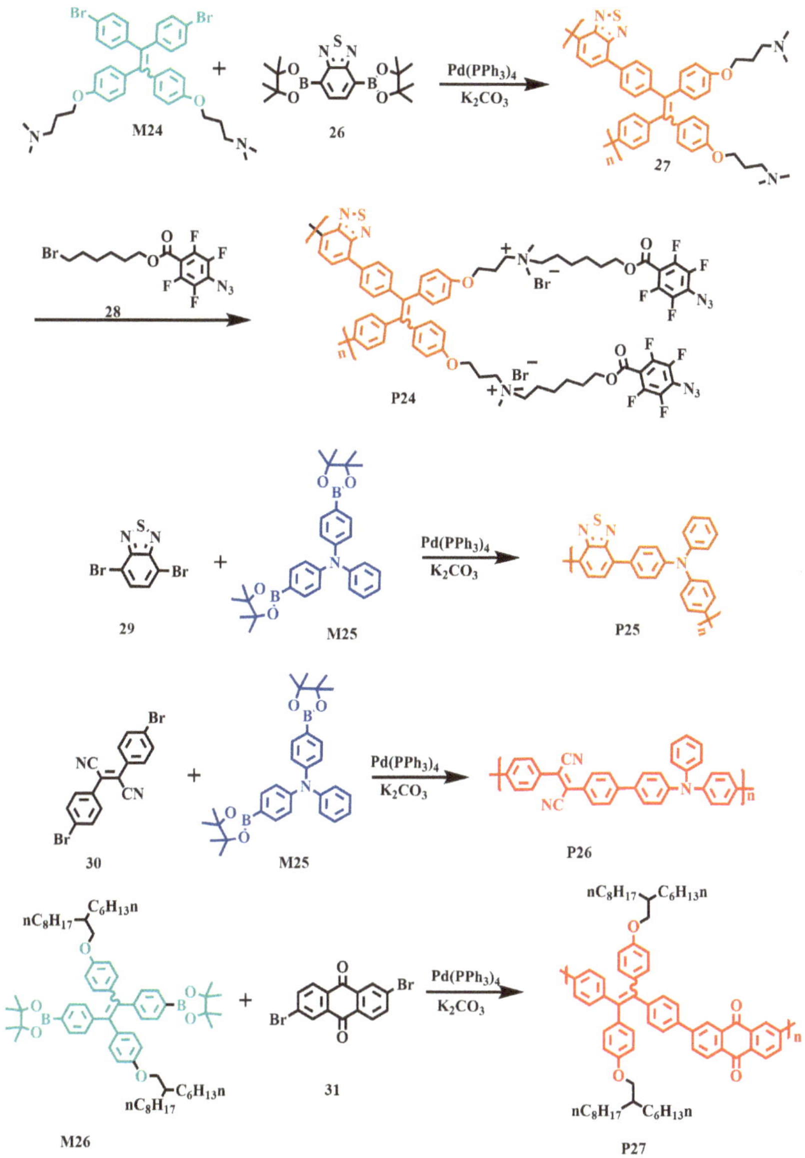

Scheme 5 Synthetic route of AIE active polymers **P24–P27** via condensation polymerization.

P25–P27 were reported for photodynamic therapy (Scheme 5). In particular, triphenylamine, TPE derivatives were reacted with benzothiadiazole, anthraquinone and 2,3-bis(4-bromophenyl)fumaronitrile using palladium tetrakis(triphenylphosphine) as a catalyst to afford the copolymers **P25–P27**.

The strong donor-acceptor copolymer structure leads to small ΔE_{ST} and enhanced intersystem crossing resulting in efficient photodynamic therapy.[22]

1.2.2.2 Knoevenagel reaction

The Knoevenagel reaction is a nucleophilic addition of carbonyl compounds with the acidic hydrogen-containing molecules to form α-β unsaturated compounds. Knoevenagel condensation is another well-known approach for designing AIE active conjugated polymers. In this polymerization method, two monomers get connected via a C=C bond. For example, triphenylamine (TPA) derivative containing dialdehyde and *p*-xylylene dicyanide were reacted under Knoevenagel reaction condition to prepare AIE active polymer **P28** (Scheme 6). The presence of donor diphenyl amino and acceptor cyano moiety at the polymer backbone leads to the efficient intrachain charge transfer. The **P28** displayed weak emission in THF, whereas its fluorescence intensity at 700 nm is enhanced in an aggregated state.[23]

1.2.2.3 Stille reaction

The Stille coupling is carried out between stannanes and halides to form C—C bonds. Through Stille coupling, generally high molecular weight polymers are obtained due to the higher reactivity of stannanes monomers than boronic acid or esters involved in the Suzuki reaction. Recently, a series of AIE active conjugated polymers **P29–P32** were reported via Stille coupling reaction. The integration of planar and twisted monomers in the polymer backbone was controlled to get high brightness NIR emitting AIE active polymer (Scheme 6). Among **P29–P32**, the **P32** polymer displayed high brightness NIR emission and utilized for fluorescence-guided cancer surgery.[24]

1.2.3 Post functionalization

An effective strategy of designing AIE active polymer is to attach the AIE active moiety at the side chain of the various naturally occurring or commercial polymers. For example, an AIE active polymer was prepared by functionalization of poly(4-vinylpyridine) with 3-(bromomethyl)-perylene. The resulting polymer was further quarternized with methyl iodide to from

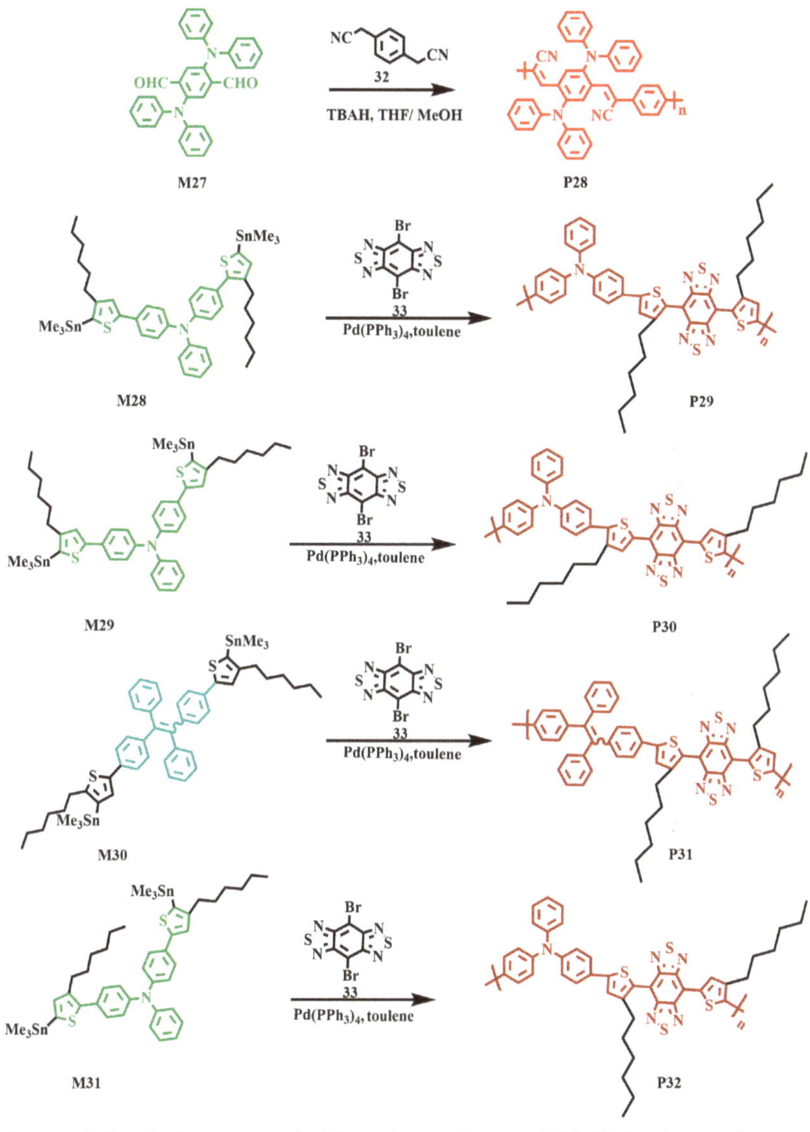

Scheme 6 Synthetic route of AIE active polymers **P28–P32** via condensation polymerization.

water soluble cationic polyelectrolyte **P33** (Scheme 7). The polyelectrolyte **P33** is weakly emissive in water (ϕ = 0.014) but emits intense fluorescence in THF or aggregate state (ϕ = 0.2).[25] Similarly, AIE active salicylaldazine-based polymer **P34** has also been reported by the successive addition of

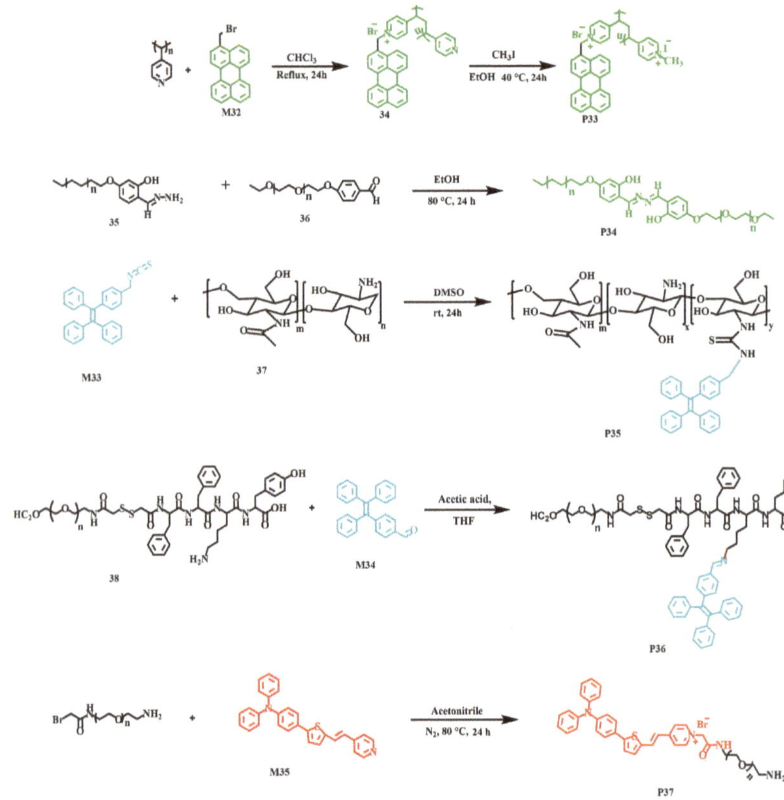

Scheme 7 Synthetic route of AIE active polymers **P33–P37** via post functionalization.

hydrazine in the mixture of 2-hydroxy, 4-mPEG-oxy benzene aldehyde and 2-hydroxy, 4-palmityloxy benzene aldehyde (Scheme 7). The **P34** forms highly fluorescent micelle in aqueous solution ($\phi = 0.14$) due to its amphiphilic property.[26] Later on a biopolymer-based AIE active polymer was reported for long-term cellular tracking. The AIE active polymer **P35** was prepared by post functionalization of biopolymer chitosan with TPE substituted isocyanate (Scheme 7). The **P35** polymer showed 36,200-fold higher PL intensity in water compared to the THF solution. It is due to the aggregation of **P35** in water, which blocks the intramolecular motion and force it to relax radiatively.[27] In another work, aldehyde containing TPE was linked with polyethylene glycol-block-peptide through nucleophilic addition reaction. The polymer **P36** forms nanoparticle in water with spherical morphology of size 100 nm (Scheme 7). The polymer **P36** was loaded with the anticancer drug DOX and form FRET pair with it. The drug release was detected by change in FRET signal.[28] Poly(ethylene glycol)-amine based AIE active polymer **P37** was constructed via post functionalization reaction of brominated poly(ethylene glycol)-amine with pyridine functionalized AIE active molecule (Scheme 7). The polyelectrolyte **P37** exhibit water solubility and solid state quantum yield of 10% compared to the aqueous solution (0.5%). The **P37** polymer was coated on the surface of 2D black phosphor to enhance the photosensitization efficiency for theranostics application.[29]

1.3 Classification of AIE active polymer

Polymers can be categorized based on structure, source, molecular force, and polymerization mode. In recent years, several AIE active polymers based on different AIE active units such as TPE, TPA, silole, phenothiazine, and even non-luminogen molecules has been explored. In this section, AIE active polymers will be classified based on polymerization mode, i.e., Addition polymers and Condensation polymers. Specifically, those AIE active polymers are discussed, which have been assessed for biological applications.

1.3.1 Addition polymers

Additional polymers are prepared by repeated inclusion of monomer containing double or triple bond. Many AIE active addition polymers were obtained starting from the vinyl group-containing AIE active monomer utilizing the well-known RAFT polymerization technique. The resulting polymer is hydrophobic and weakly emits in organic solvent. However, it forms highly emissive nanoaggregates in water due to restriction of intramolecular motion, forcing excited state chromophore to decay radiatively. Additionally, they demonstrates good dispersibility, biocompatibility and stability, making them potential candidates for biosensor and bioimaging applications. In recent report, an AIE active polymer **P1** was prepared via addition polymerization technique using vinyl substituted TPE monomer for the detection of glucose (Scheme 1). The polymer form fluorescent micelle in water with emission peak centered at 480 nm. However in presence of glucose, glucose oxidase and iodide, its fluorescence gets quenching due to the generation of I_2 in the system via cascade enzymatic and chemical oxidations. Similarly, another addition AIE glycopolymer **P2** was reported for the imaging of human lung carcinoma cells (Scheme 1). The polymer was synthetized by repeated addition reaction of vinyl substituted phenothiazine monomer. Likewise, polymer **P3** and **P33** are the examples of addition AIE active polymers prepared either direct polymerization of vinyl group containing AIEgen or post functionalized of polyethylene derivatives with the AIEgens (Schemes 2 and 7).

1.3.2 Condensation polymers

Condensation polymers are prepared by the repeated intermolecular condensation reaction of two different trifunctional or bifunctional monomers with the loss of small molecule. The two monomers involved in condensation polymerization can be AIEgens or one AIEgens and another conventional monomer. Through condensation polymerization, both non-conjugated and AIE active conjugated polymers can be formed. For example, an AIE active thermogelling polyurethane **P4** was constructed via condensation polymerization (Scheme 3). The **P4** forms weakly emissive self-assembled micellar structure at Critical Micellar concentration (CMC) and highly fluorescent gel above CMC. Moreover, several AIE active conjugated polymers were explored by means of polycondensation reactions. Such polymers enjoys advantages in terms of unique optoelectronic properties, mechanical flexibility and synthetic versatility over the small molecule AIEgens. The efficient exciton delocalization both between and along the polymer chain is responsible for its unique practical properties and functionalities. Additionally, such AIE active conjugated polymers offers supersenstivity and amplified response in chemo and biosensing applications. For instance, an AIE active copolymer of fluorene and sulfide-substituted boron diiminate, **P5** was designed for H_2O_2 detection (Scheme 3). A turn on response was observed on H_2O_2 addition as it oxide sulfide group of boron immides to the sulfoxide. Similarly, they are also explored for cell imaging, long term cellular tracking and theranostics application due to their high brightness, low cytotoxicity, photostability and tunable emission. For instance, polymers **P8**, **P9**, **P10–P18**, **P29–P32** are the examples of condensation polymers which has been utilized for cell imaging application. The condensation polymers **P19–P27** are explored for drug delivery and cancer theranostics application (Schemes 3, 4 and 6).

2. Diagnostics with AIE active polymer

2.1 Detection of biomolecules

Detection of biomolecules such as DNA, proteins, etc. and other biochemical compounds such as glucose, amino acids, etc. existing in the biological systems has been of great importance. Their abnormal levels in the body suggest the malfunction in the body system's activity. For example, glucose is a well-known biomolecule as it acts as fuel for biological processes. Its abnormal levels in urine or human blood indicate biological disorder and relate to a particular disease.[30–32] However, there are very selective reports available to detect them, which utilize particularly a AIE active polymer sensor. Majority of AIE active polymer sensors consist of TPE units which have recently gained attention because of their interesting fluorescent features such as they tend to emit only in aggregated form, while in solubilized form they are weakly fluorescent or non-fluorescent known as AIE. Therefore, many researchers utilized this moiety in the non AIE active polymeric system to acquire AIE phenomena in the polymeric system. One such incorporation was performed to synthesize AIE active polymer **P1** via RAFT polymer method by incorporation of TPE moiety in the side chains of the main backbone of the **P1** (Scheme 1).[11] This cationic polymer has hydrophobic TPE, which makes it weakly fluorescent in organic solvent and highly fluorescent in polar solvent such as water due to AIE. Thus, **P1** forms fluorescent polymeric micelles in the aqueous solution and emits strong fluorescence at 480 nm (excitation = 334 nm), which gradually quenches in the presence of I_2 as it forms a static complex with the hydrophobic TPE moieties. Furthermore, utilizing cascade enzymatic oxidation of glucose by an enzyme glucose oxidase (GO_x), hydrogen peroxide (H_2O_2) was produced as one of the products which can easily oxidize I^- ions to I_2. Hence, by taking advantage of these chemical reaction and cascade enzymatic reaction, detection of glucose was demonstrated selectively when compared to other carbohydrates such as D-mannose, D-galactose and D-fructose. However, these results can be affected in the presence of higher amount of ascorbic acids, which can be stated its one of the drawbacks. On the other hand, this method is quite applicable on real human blood samples after some preparatory steps.[11]

In addition, fluorescence probes for detection of DNA hybridization have also been accomplished via a AIE based perylene-functionalized polymer. The synthesis of a perylene-functionalized polycation **P33**, which possess poly(4-vinylpyridine) as main polymeric backbone chain and having cationic pyridinium ion in the side chain contained perylene moiety was reported, which behave as a AIE active material (Scheme 5 and Fig. 1A).[25] **P33** is soluble in water and non-fluorescent, and on decreasing the water fraction by incorporating the THF solvent its fluorescence enhances significantly, while its AIE behavior along with the excimer peak of perylene appearing at ~500–600 nm could also be seen. DNA are negatively charged, therefore tend to have effective binding with cationic species such as **P33**. Hence upon

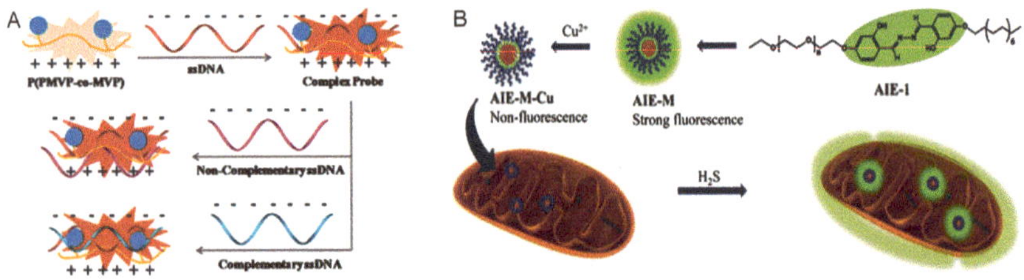

Fig. 1 (A) Conceptual illustration of the formation of the complex probe and label-free DNA and detection based on the AIE active polymer **P33**. and (B) Intercalation effects of micelles of polymer **P34** and its graphical representation for H_2S sensing. Panel A: *Adapted with the permission from Wang G, Zhang R, Xu C, Zhou R, Dong J, Bai H, Zhan X. Fluorescence detection of DNA hybridization based on the aggregation-induced emission of a perylene-functionalized polymer.* ACS Appl Mater Interfaces. *2014;6:11136–11141;* Panel B: *Adapted with the permission from Liu L, Wu B, Yu P, Zhuo R-X, Huang S-W. Sub-20 nm nontoxic aggregation-induced emission micellar fluorescent light-up probe for highly specific and sensitive mitochondrial imaging of hydrogen sulfide.* Polym Chem. *2015;6:5185–5189.*

addition of the negatively charged ssDNA1 (5′-CAA GTA GAA TGT ATG TGC-3′) at pH 7.4, it forms a complex with probe polymer, and the increment in the fluorescence intensity was observed due to AIE effect. Now addition of complementary DNA, i.e., ssDNA2 (5′-GCA CAT ACA TTC TAC TTG-3′) caused very minor change in the fluorescence. Whereas addition of the noncomplementary ssDNA, i.e., ssDNAt (5′-TTT TTT TTT TTT TTT TTT-3′) caused a significant increment in the fluorescence of the complexed probe. Thus, this AIE active polymer **P33** can distinguish the complementary and noncomplementary DNA depending on the change in the fluorescence of the complex probe. The change in the fluorescence in case of noncomplementary DNA was attributed to the AIE property of the **P33** while the minor change in case of the complementary DNA was due to intercalation of the chromophore perylene into the duplex.[33,34]

2.2 Detection of diseases related biomarkers

Hydrogen sulfide (H_2S) is a gaseous molecule having unpleasant smell like that of rotten eggs, and recently recognized as an important molecule in biological systems, referred as third gaseous signal molecule, besides carbon monoxide and nitric oxide.[35–37] Previously, various fluorescent sensors have been developed to detect it, especially its imaging in various organelle such as mitochondria, lysosomes, etc. In this regard, a fluorescence light-up AIE based polymeric micelles (AIE-M) having salicylaldazine moiety was designed, which had rotating N—N bond in good solvent and thus responsible for its AIE behavior in an aggregated state.[26] **P34** is highly soluble in DMSO and is very weakly emissive (Scheme 7). Its highly fluorescent micelles (AIE-M) were prepared by dropwise addition of its DMSO solution into water under ultrasonic treatment. These fluorescent micelles AIE-M formed a coordination complex (AIE-M-Cu) with Cu^{2+}, and the fluorescence of the AIE-M quenched significantly. Further, among various analytes namely chloride, carbonate, nitrate, phosphate, thiosulfate, cysteine and GSH, the non-fluorescent AIE-M-Cu showed only turn-on behavior in the presence of S^{2-} ions. This high selectivity was achieved due to the selective CuS precipitation. Similar fluorescence behavior was also seen in mitochondrial imaging of H_2S via AIE-M-Cu probe (Fig. 1B).

Hydrogen peroxide (H_2O_2) is one of the reactive oxygen species (ROS) and produced under oxidative metabolism. It acts as a mediator and is involved in many cell signaling pathways.[38–40] Its concentration in normal *in vivo* conditions remains $\sim 10^{-7}$ M, while in case of inflammation condition concentration rises above 10^{-6} M.[38] Therefore, it can be considered as one of the biomarkers for the inflammatory diseases namely atherosclerosis, arthritis, cancer and so on. Thus, its sensitive detection is highly desirable. In this direction, AIE active polymer-based films for the fluorescence turn-on detection of hydrogen peroxide have been developed.[15] Three polymeric derivatives **P5–P7** of various monomers of boron diiminates (M5, M6 and M7) with a functionalized fluorene derivative via palladium catalyzed Suzuki-Miyaura coupling polymerization method were reported in high yields (Scheme 3). These polymers displayed good solubility in organic solvents such as THF, chloroform, and dichloromethane with a good film forming property. The synthesized polymers have subtle luminescence in THF solution with quantum yield, i.e., $\phi_{PL} < 0.01$, while in films have $\phi_{PL} = 0.03–0.09$, indicating the AIE behavior in aggregated films. The quantum yield of **P5** and **P6** were higher than **P7**, indicating involvement of significant charge transfer interaction between boron diiminates and fluorene units in oxidized state. For sensing of H_2O_2, films of polymer **P7** were taken for the oxidation in presence of 30 wt% H_2O_2 solution and in acidic condition along with Na_2WO_4. There was increment in the quantum yield of the polymeric film from 0.034 to 0.087 after 240 min. This increment of fluorescence in the **P7** film suggested the transformation into sulfoxide group by H_2O_2. Using this concept, various portable films can be achieved not limited to just chemosensors, but also for biological applications.

2.3 Detection of microbes and antibacterial application

Human health is prone to many infectious diseases, and bacterial infections are one among them. In recent times more prevalent water-borne pathogens have been observed due to increased environmental pollution and global climate change.[41,42] Therefore, detection as well as combating bacteria is highly desirable. Previously, many methods have already been established for this purpose such as polymerase chain reaction (PCR) and standard plate count (SPC), but these methods have certain drawbacks like requiring expert handler and long incubation time.[43,44] Therefore, detection of these bacterial samples without any complicated protocol are more preferred. In this regard various platforms for the detection of bacteria have been achieved based on various methods such as colorimetric, fluorometric and magnetic resonance imaging (MRI). In this direction a TPE-centered cationic star amphiphilic copolymer, i.e., TPE-star-P(DMA-co-BMA-co-Gd), via ATRP method having a TPE core, for the detection and inhibition of bacteria based on MRI and fluorescent signals has been reported.[45] The side arms of the copolymers are amphiphilic/cationic P(DMA-co-BMA-co-Gd) arms, where Gd, BMA and DMA are T1-type MRI contrast agent, butyl methylacrylate and 2-(dimethylamino)ethyl methacrylate, respectively. These cationic side chains, interact with the negatively charged membrane of bacteria resulting in electrostatic complexation, which restricts the intramolecular rotation of the TPE core moiety. As a result, this gives rise to a turn-on fluorescence response along with enhancement of magnetic resonance longitudinal relaxation rate. Apart from the sensing, the star copolymers bear antibacterial behavior against both Gram-positive (*S. aureus*) and Gram-negative (*E. coli* and *P. aeruginosa*) bacteria stains.[46]

3. Imaging with AIE active polymer

3.1 Imaging with polymer

Bioimaging includes a wide range of non-invasive imaging that enables visualization of macroscopic and microscopic levels of biological and anatomical processes distinctly. It has broad applications from cellular to detail transmitting and processing information of entire body. For clinical diagnosis and research work, visible light microscopy is a well-established technique and plays a very significant role.[47] The AIE conjugated polymers or fluorescent dyes have been broadly used to visualize biological process and tissue architecture.[48] Fluorescent dyes are also used in flow cytometry to determine the cell viability and population. Currently, bioimaging techniques are based on ionizing or non-ionize electromagnetic radiation.[49] The X-ray computed tomography (CT) has been used since 1972 and Positron emission tomography (PET) was used as a replacement of the CT technique).[50] Changing into non-ionizing technique, the magnetic resonance imaging (MRI), ultrasound (sonography) has emerged as a potent tool for imaging in the last 30 years.[51]

The currently available bioimaging techniques have some limitations with regards to resolution, penetration and sensitivity. Consequently, diverse bio-probes have been developed to enhance imaging. Different fluorescent materials such as quantum dot, small molecules and AIE active conjugated polymers have been designed. The AIE active conjugated polymers has attracted immense attraction due to its tailored fluorescence spectrum, which can be used in photoacoustic and fluorescence imaging. Further AIE active conjugated polymers are photostable, simple synthesis method, ease in modification, higher quantum yield, and superior sensitivity compared to small molecules.[52]

AIE active polymer attributes a novel class of fluorescence materials that exhibit enhanced emission when aggregated. Due to the excellent fluorescence response toward biomolecules, AIE active polymers are promising materials for bioimaging. AIE active polymers enjoy intense emission and superior stability they also have outstanding biocompatibility, good water dispersibility and solubility.

The AIE active conjugated polymers show photoluminescence from visible to near-infrared region (NIR) (Fig. 2) based upon its optical tuning. Hence AIE active polymers have higher potential for *in vivo*, *in vitro* imaging applications.[53]

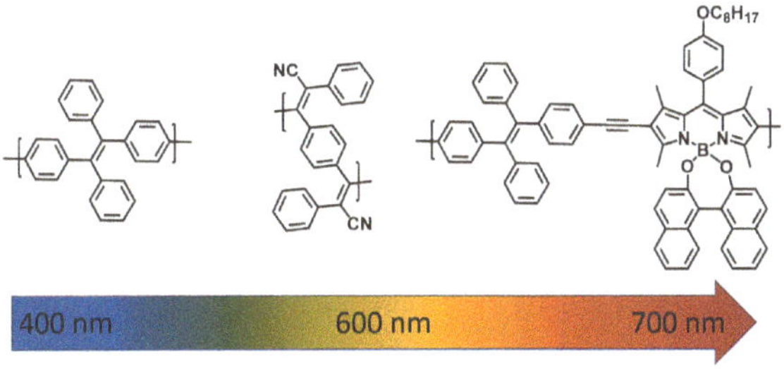

Fig. 2 Color tunable AIE active conjugated polymers with their corresponding emission regions (from the blue region to the NIR region).

3.2 AIE active polymer for *in-vitro* imaging

Self-assembly to nanoparticles formation in an aqueous medium is a promising strategy for bioimaging applications. A one-pot synthesis of AIE active polymer was reported for the first time where the polymer was synthesized by using a free radical polymerization method. AIE monomer PhE, itaconic anhydride and glucosamine hydrochloride was polymerized to obtain a glycopolymer **P2** (Scheme 1). The polymer was covalently attached with the AIE dye hence it prevents the leakage of dyes and furthermore, the polymer surface was covered with the glycosyl groups this makes the higher potential and superior biocompatibility for further biological applications. This polymer is amphiphilic in nature and prone to form nanoparticles. It has higher water dispersibility due to its carboxyl and glycosyl groups lead to stable morphology. The cytotoxicity and biocompatibility of **P2** was performed to the human lung adenocarcinoma epithelial (A549) cells. The cell viability was found greater than 90% up to 80 $\mu g\ mL^{-1}$ concentration. The fluorescence imaging of **P2** was performed by a confocal laser scanning microscope (CLSM). **P2** uptake into cells could be uptaken by endocytosis that leads to intense orange fluorescence.[12]

The AIE active conjugated polymers have higher molecular weight and higher conjugation that generate a synergistic effect. Due to this AIE active conjugated polymers show exceptional advantage over small molecules dyes. Discrimination and detection of live cells (viable cells) is very significant for monitoring the cell activity to provide important information. For instance, TPE linked AIE active polymers were designed to discriminate dead and apoptotic cells selectively from viable cells (Fig. 3A).[16] **P8** had AIE emission and great photophysical properties, higher selectivity, biocompatibility, etc. (Scheme 3). They found the photostability of **P8** is comparable to Calcein AM, hence it is an ideal probe for bioimaging, and performed binding of **P8** with the viable, apoptotic and dead cells. **P8** labeled only viable cells with bright emission with higher sensitivity but have least affinity toward the apoptotic and dead cells. It successfully discriminated dead cells from viable cell on the basis of zeta potential and electrostatic repulsion (Fig. 3B). The negatively charged **P8** AIE active polymer with the side-chain and endocytosis both played critical roles in this selective discrimination making it an amazing tool for precise guiding cancer therapy.

Smart polymeric micelles with intense emission in NIR range are in great demand for cancer and healthy cell imaging. Recently pH and redox responsive AIE active micelles for cell imaging have been reported.

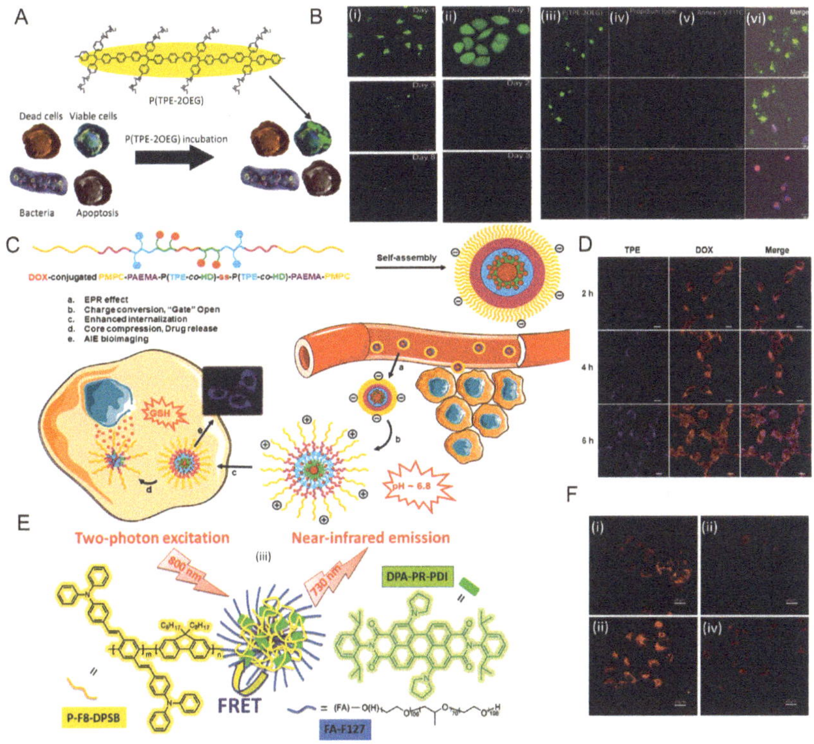

Fig. 3 (A) Pictorial representation of selective detection of viable mammalian cells over dead cells, bacteria, and apoptosis cells by **P8**. (B) CLSM images of (i) **P8** (ii) Calcein AM in HeLa cells incubated for 1, 2, 3 days and (iii) **P8** (after treated by (1) 0 (2) 500 (3) 1000 μM H_2O_2 for 6 h and then stained by (iv) PI and (v) annexin V-FITC (vi) merged images of B(i)–B(v). (C) Illustration of pH and redox responsive polymer micelles drug delivery and imaging (D) CLSM pictures of 4T1 cells with polymeric micelles for various times. (E) Schematic representation for synthesis of hybrid NPs and NIR fluorescence emission by two photon excitations. (F) Deep penetration imaging of Hela cells of hybrid NPs (8 μg mL^{-1}) incubation (i) cell fluorescence images with internalized hybrid NPs (ii) hybrid NPs with encapsulation of FA-F127 fluorescence was intensified (iii) and (iv) with internalized F127-encapsuled and FA-F127-encapsuled hybrid NPs, respectively, in the presence of mock tissue with 900-μm thickness. Panel B: *Adapted with the permission from Hu R, Zhou T, Li B, Kwok RT, Shen J, Qin A, Tang B.Z Selective viable cell discrimination by a conjugated polymer featuring aggregation-induced emission characteristic.* Biomaterials. *2020;230:119658;* Panel D: *Adapted with the permission from Su X, Ma B, Hu J, Yu T, Zhuang W, Yang L, Li G, Wang Y. Dual-Responsive doxorubicin-conjugated polymeric micelles with aggregation-induced emission active bioimaging and charge conversion for cancer therapy.* Bioconjug Chem. *2018;29:4050–4061;* Panel F: *Adapted with the permission from Lv Y, Liu P, Ding H, Wu Y, Yan Y, Liu H, Wang X, Huang F, Zhao Y, Tian Z. Conjugated polymer-based hybrid nanoparticles with two-photon excitation and near-infrared emission features for fluorescence bioimaging within the biological window.* ACS Appl Mater Interfaces. *2015;7:20640–20648.*

This copolymer **P3** was self-assembled into core-shell like micelles, and conjugated via pH cleavable linkage with doxorubicin (Scheme 2). TPE dye located in the center of the core, made these micelles capable for bio-imaging with great sensitivity. Furthermore, under tumor (acidic conditions) the pH-controlled "gate" PAEMA opened up allowing faster drug release. (Fig. 3C). The drug delivery in intracellular can be traced by turn-on response form AIE micelles. Therefore, this polymer was incubated with the mouse mammary carcinoma 4T1 cells for different time duration and found that DOX drug efficiently released from micelles and inhibited tumor growth (Fig. 3D).[13]

Fluorescent hybrid nanoparticles capable of NIR emission has attracted immerse attention. An AIE active polymer **P9** was used as an energy donor and NIR light-harvesting dye component. DPA-PR-PDI dye was used as the energy acceptor and light-emitting component for the NIR fluorescent nanoparticles (Scheme 3). This NP probe has two-photon excited fluorescence (TPEF) and Förster resonance energy transfer (FRET) that generates NIR emission capability (Fig. 3E). Furthermore, to increase selective recognition of cancer cells, the NPs were functionalized with FA-F127. Hybrid NPs cytotoxicity was performed in HELA cells to find only 5% viability that decreased to the concentration of 10 $\mu g\,mL^{-1}$. The practicability of the novel NP to the *in vitro* imaging was performed successfully with higher resolution. Thick tissue imaging could be possible with this hybrid NPs due to its higher penetration depth ability up to 12 μm (Fig. 3F).[17]

3.3 AIE active polymer for *in-vivo* imaging

NIR or IR dyes have great importance for *in vivo* imaging since they have higher spatiotemporal resolution with deep penetration ability. Most of the NIR dyes suffer from notorious ACQ effects, which limits its application in bioimaging. A signal-boosted nanophotonic AIE nanoprobe by combining NIR emitting polymer **P28** and AIE active polymer is reported (Scheme 6). Using this AIE active polymer, a NP imaging probe was constructed by incorporating two other small molecules bearing unique photonic features. First, a peroxalate (CPPO) is used as a chemical fuel generating electronic excitation energy in response to inflammatory H_2O_2. Second, a green emitting BODIPY dye is used which relays the chemically generated excitation energy to the emitter for its efficient excitation (Fig. 4A). It can be clearly seen in the chemiluminescence (CL) spectra that the BODIPY doped CLNP-PPV significantly boosted the luminescence in NIR region due to

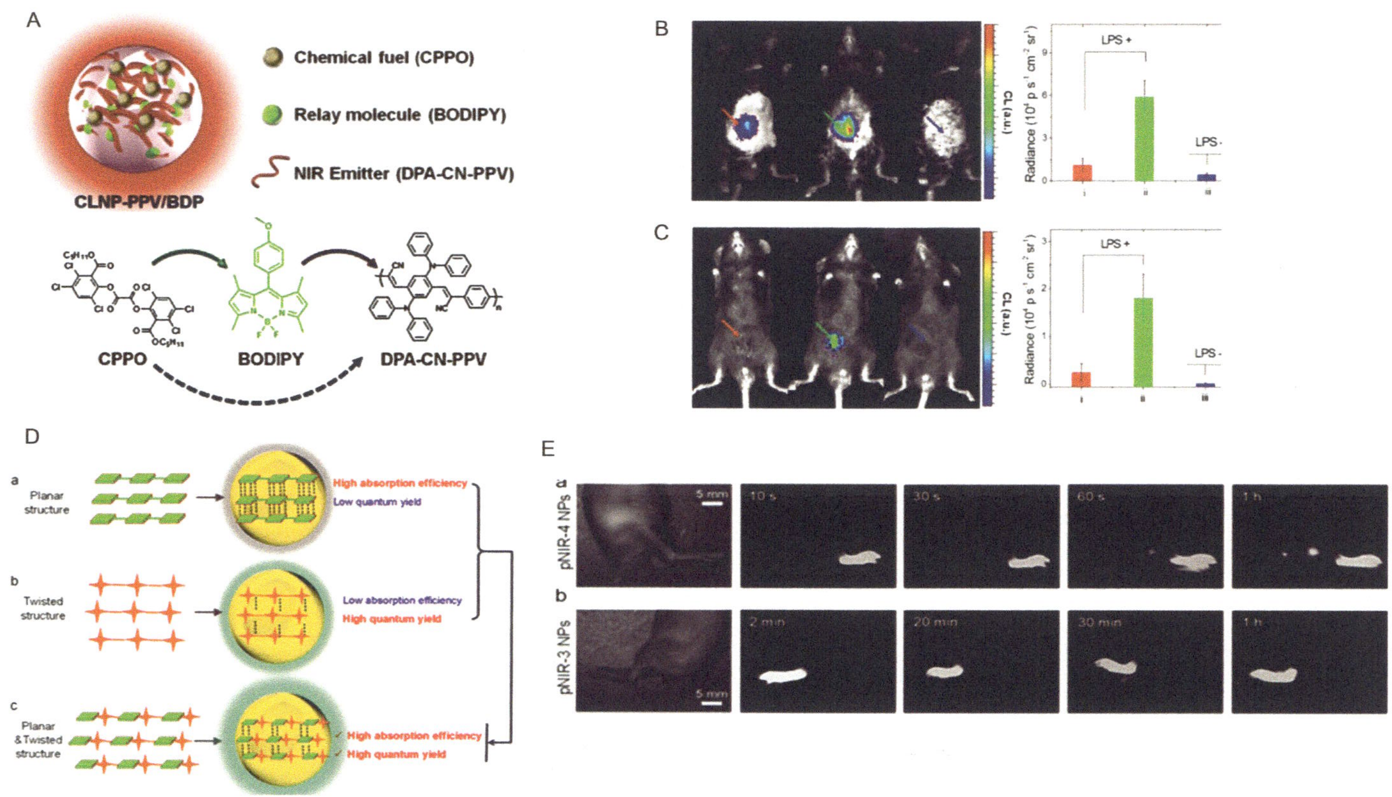

Fig. 4 See figure legend on next page.

the effective relay of energy to the AIE active polymer **P28**. This significant boosting of NIR CL was further evaluated in *in vivo*-mimicking condition. H_2O_2 responsive CL signals were imaged through various thickness of biological tissues. The penetration limit for the doped CLNP-PPV/BDP was much higher (>12 mm) than un-boosted CLNP-PPV, i.e., ~6 mm. This indicated that the NIR CL is less attenuated by the thickness of the tissue. In addition, CL imaging of inflammation in case of LPS-induced early stage of arthritis in mouse model was monitored. Only one joint of the mouse's leg was treated with LPS and then both the joints were administered with probes. The leg treated with LPS, displayed red-green signal, indicating overexpressed levels of H_2O_2 in the inflamed tissues and amplified in case of boosted probe, i.e., CLNP-PPV/ (Fig. 4B and C).[23]

The AIE active polymers with emission in the NIR-II region has improved brightness by reducing the autofluorescence and photon scattering in the tissue. Further, AIE active polymer dots **P10–P18** for NIR-IIa mouse brain imaging have also been developed (Scheme 4). By combination of an AIE molecule to increase the brightness, phenothiazine molecule to prevent self-quenching, the emission was further boosted by using non-planar molecules and tuned color by incorporating different electron affinity acceptors. The resulting AIE active polymers showed 21 times emission enhancement in aqueous medium compared to soluble state (THF), and have higher signal to background ratio. *In vivo*, non-invasive fluorescence bioimaging was performed by injecting the Pdots via the mouse tail and the mouse head with intact skull and scalp image was taken with animal imaging setup with high resolution.[18]

NIR emissive AIE active polymers are very interesting fluorogenic materials because it can be used in mammal's deep tissue imaging. These AIE active polymers were lacking in brightness in the NIR-II region.

Fig. 4 (A) CLNP-PPV/BDP nanoparticles and its intraparticle energy-relay schematic representation. (B) Peritonitis in hair-shaved *in vivo* imaging and intensity plots (right) (C) unshaved mice ($n=3$) image taken after at 1 min. (D) Pictorial representation of novel design for bright emissive dye. NIR-II fluorescence pictures of the lymphatic system of live mice by (E) pNIR-4 and pNIR-3 NPs. Panel C: *Adapted with the permission from Seo YH, Singh A, Cho H-J, Kim Y, Heo J, Lim C-K, Park SY, Jang W-D, Kim S. Rational design for enhancing inflammation-responsive in vivo chemiluminescence via nanophotonic energy relay to near-infrared AIE-active conjugated polymer.* Biomaterials. *2016;84: 111–118;* Panel E: *Adapted with the permission from Liu S, Ou H, Li Y, Zhang H, Liu J, Lu X, Kwok RT, Lam JW, Ding D, Tang BZ. Planar and twisted molecular structure leads to the high brightness of semiconducting polymer nanoparticles for NIR-IIa fluorescence imaging.* J Am Chem Soc. *2020;142:15146–15156.*

Such semiconducting AIE active polymers, **P29–P30** are explored to get bright emission in NIR-II region (Scheme 6). In the first step, by using phenothiazine, AIE was generated and the nonradiative decay pathway was reduced while in the second step, steric hindrance was created to further minimize the quenching effect by π-π stacking (Fig. 4D). Semiconducting polymer dots (Pdots) have tunable optical properties and hence they can be extensively used in biosensing and bioimaging. The feasibility of **P30** investigated in the mouse brain and in the hindlimb vasculatures by intravenous injection were extensively monitored. Their potential utility in the clinical samples was performed by lymphatic imaging. pNIR-4 NPs brightness outperformed several classical NIR-II dyes and inspired the development of very bright NIR-II dyes (Fig. 4E).[24]

4. Biomedical application of AIE active polymer

4.1 AIE active polymer for drug screening and drug delivery

Recently, Drug delivery systems (DDSs) based on AIE active polymer have been extensively implemented as drug delivery carriers to deliver the drugs to tumors for cancer treatment due to their prolonged blood circulation and low cytotoxicity.[54] The AIE DDSs have been developed for drug delivery and bioimaging in one platform with high sensitivity in the condensed-state compared with the traditional fluorescent DDSs which impedes the bioimaging performances due to its ACQ behavior.[55–57] In addition, the polymer based materials often shows decent drug loading capacity, high stability and good work ability. The therapeutic efficacy of drug is crucially rely on the effective drug concentrations in the targeted tumors and cancerious cells. The novel DDSs exhibit simultaneously drug distribution, drug release/activation characteristics and predict the therapeutic responses. Upon intravenous injection, the drugs accumulate in the targeted tumor tissues through active or passive targeting, then pass through the tumor extracellular matrix (ECM).[58] After binding to the cells, they cross the cell membrane to enter the targeted cancerous cells and then release the drugs.[59] Numerous imaging-guided AIE active polymer as DDSs have been reported so far.[60–70]

Hydrophilic poly ethylene glycol (PEG), hydrophobic poly propylene glycol (PPG) and TPE based AIE thermogelling copolymer **P4** have been developed for monitoring the slow and localized *in vivo* drug release at the tumor site, which resulted the high efficiency at reducing the tumor size (Scheme 3). This AIE thermogel displayed exponentially higher emission in the aggregated-state than the conventional AIE micelles due to the enhanced restricted intramolecular rotation (RIR) process in the closed-packed gel. Importantly, the emission from the **P4** gel depot was detectable up to 18 days. After being internalized by the A549 tumor cells via an energy-dependent endocytosis, the thermogel was transported into the lysosome and doxorubicin (DOX) was detached from the AIE micellar aggregates due to the acidic environment of the lysome. The concentration of DOX was increased into the nucleus and simultaneously the AIE effect of TPE was recovered. This biocompatible, self-indicating AIE thermogel can load up to 50 $\mu g\ mL^{-1}$ DOX drug. Therefore, this AIE **P4** gel was an excellent candidate for *in vivo* drug concentration detection and localized delivery of the chemotherapeutic drug DOX. Most importantly, the localized delivery of the chemotherapeutic drug executed more efficiency in the tumor treatment and the reduction of the reappearance of the tumor. In addition, slow release of the drug result low cytoxicity.[14]

A visible light-activatable DOX-loaded Pt(IV) prodrug and TPE functionalized AIE copolymer nanoparticle (PtAIECP@DOX NP) system for dual-drug monitoring and combinational anticancer efficacy through the fluorescence "turn-on" mechanism both *in vitro* and *in vivo* has been explored carefully. After applying visible light, the self-assembled structure was dissociated and the Pt(IV) prodrug activates to Pt(II) and simultaneously encapsulated DOX was released intracellularly, which then entered into the cell nucleus. After releasing DOX, the fluorescence intensity of TPE was restored and FRET between the TPE and DOX was destroyed gradually. The IC50 value of PtAIECP@DOX NP (5.66 μm/Pt, 0.2 μg/DOX) was much lower than that of DOX (1.09 μg/DOX) and the anti-cancer drug cisplatin (8.9 μm/Pt). These NPs exhibited ca. 3.5-fold higher accumulation in cancer cells than the cisplatin (Fig. 5A and B).[71]

Recently three AIE active amphiphilic copolymers, **P19–P21** have been reported, using the TPE moiety and by adjusting the donor-acceptor electronic structure (Scheme 4). Notably, their fluorescence color was tuned from yellow to deep red by simply changing their donor acceptor structure. After loading the anticancer drug paclitaxel (PTX), two AIE active polymers **P19** and **P20** could be prepared into polymer dots (**P19** and **P20** dots) via the nanoprecipitation method for applying in self-illuminating cancer therapy. The **P20** dots remained in the cell cytoplasm whereas the **P19** dots gradually accumulated in the cell nucleus from the cytoplasm. The PTX-loaded **P19** and **P20** dots reveal their locations by changing their yellow or red emission and exhibit almost two times *in vivo* antitumor efficacy,

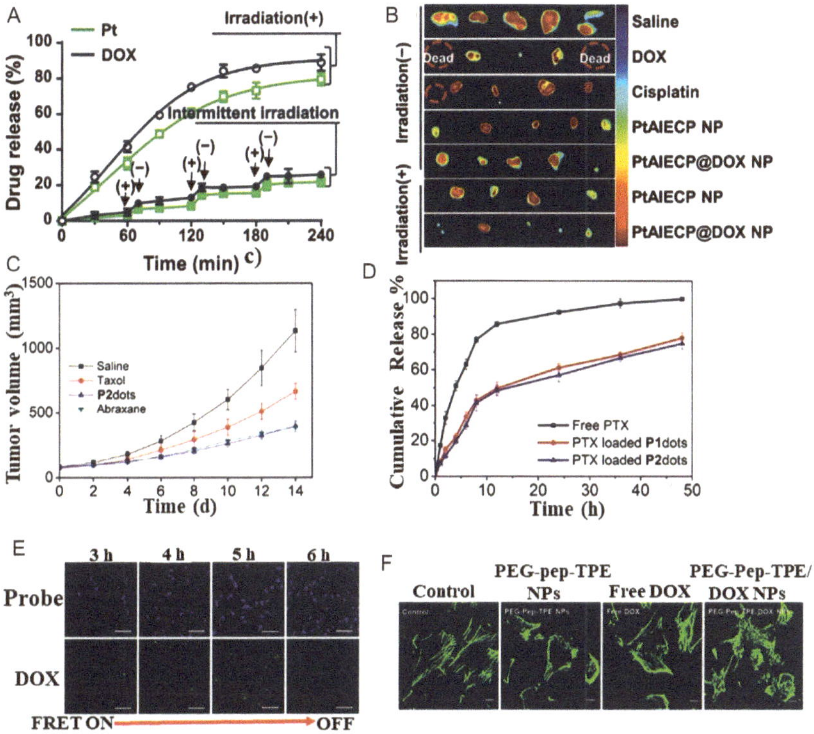

Fig. 5 (A) Pt and DOX release diagram of PtAIECP@DOX NP in different irradiation conditions. Light source: visible light (500 nm, 20 mW/cm^2). (B) *Ex vivo* red fluorescence protein (RFP) fluorescence imaging of the tumor. (C) Inhibition of tumor volume of the nude mice after intravenous injection of the saline, Taxol, PTX-loaded **P20** dots, and Abraxane for 14 days. (D) Cumulative release diagram of free PTX and PTX-loaded **P19** and **P20** dots over the course of 48 h in PBS (pH 7.4). (E, F) Confocal images of A549 cancer cells over the time after incubation with the **P36**/DOX NPs for 3 h at a dose of 10 μg mL^{-1} DOX (left) and of cell actin filaments after the treatment with PBS control, **P36** NPs, free DOX and **P36**/DOX NPs for 12 h (right). Panel B: *Adapted with the permission from Wu P, Wang X, Wang Z, Ma W, Guo J, Chen J, Yu Z, Li J, Zhou D. Light-activatable prodrug and AIEgen copolymer nanoparticle for dual-drug monitoring and combination therapy.* ACS Appl Mater Interfaces. *2019;11:18691–18700;* Panel D: *Adapted with the permission from Wang Z, Wang C, Gan Q, Cao Y, Yuan H, Hua D. Donor–acceptor-type conjugated polymer-based multicolored drug carriers with tunable aggregation-induced emission behavior for self-illuminating cancer therapy.* ACS Appl Mater Interfaces. *2019;11: 41853–41861;* Panel E and F: *Adapted with the permission from Wang T-T, Wei Q-C, Zhang Z-T, Lin M-T, Chen J-J, Zhou Y, Guo N-N, Zhong X-C, Xu W-H, Liu Z-X. AIE/FRET-based versatile PEG-Pep-TPE/DOX nanoparticles for cancer therapy and real-time drug release monitoring.* Biomater Sci. *2020;8:118–124.*

high biocompatibility and obviously tumor-targeting characteristics compared to the commercial available anticancer drug Taxol. Additionally, **P2**-dots displayed similar *in vivo* antitumor efficacy and biocompatibility compared to the non-emissive commercially available PTX-loaded DDS Abraxane. The drug loading content (DLC) (as a function of weight) of **P20** dots and **P19** dots were of 10.97 and 10.57%, respectively. This strategy established the significant application of a donor-acceptor type multi colored AIE active systems in the design of self-illuminating drug carriers (Fig. 5C and D).[19]

An AIE active amphiphilic copolymer **P36** with the hydrophobic D-peptide-TPE (Pep-TPE) and hydrophilic PEG was recently reported (Scheme 7). The DOX-loaded **P36** was self-assembled into nanoparticles (NPs) (**P36**/DOX NPs) in aqueous solutions. The pH/glutathione (GSH)-responsive **P36**/DOX NPs was applied for combinational cancer imaging and the real-time monitoring the drug release process based on the FRET coupled with the AIE luminogens via fluorescence "turn-on" effect. This **P36**/DOX NPs can deliver the drug into the A549 cancer cells at a dose of 10 $\mu g\, mL^{-1}$ DOX. The IC50 value and *in vitro* drug release efficiency of **P36** + DOX NP were 0.3735 $\mu g\, mL^{-1}$ and 93.59%. Herein, the fluorescent anticancer drug DOX acts as an acceptor to become a FRET pair with AIE luminogens. When the entrapped drug DOX was released from the nanoparticles, the FRET was gradually disappeared and the intensed blue emission from the TPE part was restored. Then the free DOX entered into the cell nucleus. In addition, the self-assembling peptide NPs impede the formation of actin filaments in the cytoplasm and even damage the existing actin in the cancer cells which resulted in the programed cell death (Fig. 5E and F).[28]

4.2 AIE active polymer for photodynamic therapy

PDT is an emerging non-invasive therapeutic modality for disease treatment, which involves the use of light, photosensitizer (PS) and endogenous molecular oxygen to kill the microbes or cancer cells.[72] PSs with AIE or aggregation-induced enhanced emission (AIEE) display strong photosensitization ability and enhanced emission in condensed-state or aggregated-state compared with the traditional PSs, which produce lower singlet oxygen and exhibit quenched fluorescence in the condensed-state. It is also witnessed that the conjugated AIE active polymer PSs are much more efficient in singlet oxygen ($^1O_2$) production than their small molecule AIE PSs due to their enhanced inter system crossing (ISC) process and improved light-harvesting ability.[73] Generally, the light in the 600–950 nm wavelength window is an optimal in PDT, which is called the therapeutic window and most of the cases white light is used for the PS irradiation because of the low cost, large illumination field and simple construction over the laser-sources.[74] In some conditions, PDT is applied along with the radiotherapy or chemotherapy to increase the therapeutic efficacy. Generally, an AIE PS is incorporated inside the targeted cell, which produce cytotoxic ROS upon light irradiation that are responsible for the cell death via apoptosis or autophagy.[75] It is observed that the PDT is highly executed in some organelles including mitochondria, lysosome, cellular membrane,[76] because these subcellular organelles have crucial roles in manipulating cellular processes.

The PDT-mediated cytotoxicity is depend on the photochemical reactions between the excited state of PS and molecular oxygen or cellular substrates. Upon light irradiation, the PSs absorbed photons and the electrons from singlet ground state (S_0) are excited to the higher-energy state (S_n). Then the excited PS can undergoes lowest excited singlet state $(S_1) \rightarrow S_0$ transitions by releasing radiative light energy as fluorescence or non-radiative heat energy through the internal conversion (Fig. 6). It is shown that the smallest ΔE_{ST} (the energy difference between the S_1 and the lowest excited triplet state T_1) value is sufficient for the transition of electrons from S_1 to the T_1 state through ISC process. The electronic transitions from $T_1 \rightarrow S_0$ state can undergo through non-radiative decay processes or phosphorescence. It is well established that the $T_1 \rightarrow S_0$ transition is quantum

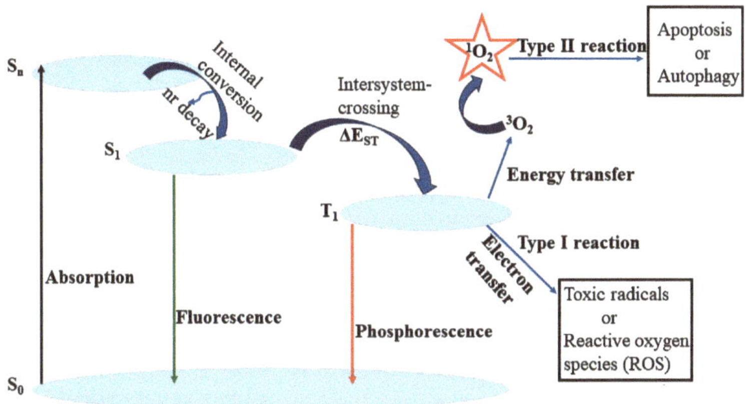

Fig. 6 Modified Jablonski diagram illustrated the principle of PDT. (nr represents the non-radiative decay.)

mechanically forbidden and T_1 state usually has a longer lifetime (microseconds to seconds). The longer lifetime of PS in T_1 state is sufficient to transfer its energy to adjacent ground state triplet oxygen ($^3O_2$), which can produce highly reactive $^1O_2$ (type II reaction). PS in T_1 state can also transfer electrons to cellular substrates (type I reaction) and eventually formed ROS and toxic radicals. Therefore, PDT could cause cancer cells or bacterial cells ablation in both types of photochemical reactions (Fig. 6).[77–79]

PSs, the key element in PDT, have been explored extensively. In addition to image-guided PDT, the AIE PSs have been also used for gene and drug delivery. The effective PSs should have the basic properties which include strong photosensitization performances upon light irradiation, high absorption coefficient at longer wavelength, good biocompatibility in dark condition and good resistance to photo bleaching. A lot of PSs based on the AIE active polymer have been reported so far.[80–83]

A water soluble NH_2-PEG-TTPy AIE active polymer and prepared BP@**P37** nanosheets by coating on the surface of 2D black phosphorous (BP) through electrostatic interactions have been developed (Scheme 7). This highly stable and excellent biocompatible hybrid BP@**P37** nanomaterial efficiently generate ROS and display strong far-red/near-infrared (FR/NIR) fluorescence and good photothermal conversion efficiency. The ROS production efficiency of BP@**P37** nanosheet was 292-fold higher than that of the commercially available PS chlorin e6 (Ce6). The superior ROS generation ability of nanosheets could be attributed to the more efficient ISC of NH_2-**P37** PEG-TTPy, resulting from both the donor-acceptor strength and heavy atom effects. It was demonstrated that BP@**P37** nanosheets were perfectly colocalized with the lysosomes after internalizing into the cells for 1 h. Then the **P37** separated from the BP nanosheets because of lysosomal acidic environment and escaped the lysosomes due to the proton sponge effect and accumulate into the mitochondria through electrostatic interactions. After 4 h incubation, upon simultaneous white light (WL) and 808 nm NIR laser light irradiation, the mitochondria disrupted because of ROS production. The remaining BP nanosheets and **P37** exhibited both ROS-induced cell apoptosis and hyperthermia in the lysosomes. The BP@**P37** nanosheets could induce autophagy (lysosomal degradation Process) also. These features significantly performaned *in vitro* and *in vivo* cancer theranostics involving NIR fluorescence and photothermal (NIR FLI-PTI) dual imaging-guided synergistic PDT and photothermal therapy (PTT). This nanosheets displayed the best tumor inhibition efficiency and the tumors were completely removed after the second day of the treatment, and no reappearance of the tumor was observed during the treatment. This BP@**P37** nanosheets not only inhibit the cancer cell proliferation but also damage the tumor vascular tissues by synergistic treatment with the PDT and PTT pathways.[29]

TPE based two AIE active polymers PSs **P22** and **P23** from its small molecule AIE PS TPEDC were developed and applied for both *in vitro* cancer cell ablation and *in vivo* liver tumor treatment of zebrafish which was induced by 2-photon excited PDT (2PE-PDT) with high efficiency (Scheme 4). Upon two-photon excitation, **P22** and **P23** dots (prepared from amphiphilic polymer TAT-SH) generates highly efficient $^1O_2$ both in aqueous solution and in cells because of enhanced ISC process and improved molecular conjugation of **P22** and **P23**. The **P23** dots displayed 548% more efficient $^1O_2$ and 651% larger 2 photon absorption (2PA) cross-section compared to the TPEDC dots. The denser energy levels of **P23** compared to **P22** and TPEDC favors the more efficient $^1O_2$ generation ability of **P23**. It was also shown that the much more smaller dihedral angles between the different units of **P23** compared to the TPEDC and **P22** increase the molecular conjugation of **P23**, which favors the yield of a larger 2PA cross-section for **P23**. Thus this AIE active polymers PSs with enhanced two-photon photosensitization provides an efficient platform for exploring highly efficient photosensitizers for 2PE-PDT (Fig. 7A and B).[20]

Further TPE functionalized AIE active polymer **P24** for the selective detection of microbes over the mammalian cells both *in vitro* and *in vivo* were also reported recently (Scheme 5). It generates ultra-strong ROS in the condensed-state upon both white light and sunlight irradiation and the ROS generation ability of this polymer is 11 and 13 times higher than that of Ce6 and its monomer. The $^1O_2$ quantum yield of polymer was calculated to be 38%, which was better than that of its monomer (29%) because of lower ΔE_{ST} value of the polymer (ΔE_{ST} for polymer was 0.881 eV and monomer 1.773 eV), which enhanced the ISC process for the polymer. This polymer was found to be mainly localized on the cell wall of the microbes especially *S. aureus* (gram positive bacteria) and distributed into the cell cytoplasm after incubation with the dead microbes. Notably, the antibacterial efficacy is even faster than that of the commercially available antibiotic cefalotin. Thus, the PTB-APFB AIE active polymer appears as a promising antibacterial agent for practical applications.[21]

Another triphenylamine (TPA) functionalized mitochondria targeting AIE active polymer **P25** was also explored for the application in anticancer therapy (Scheme 5). The $^1O_2$ quantum yield of this polymer AIEgen is

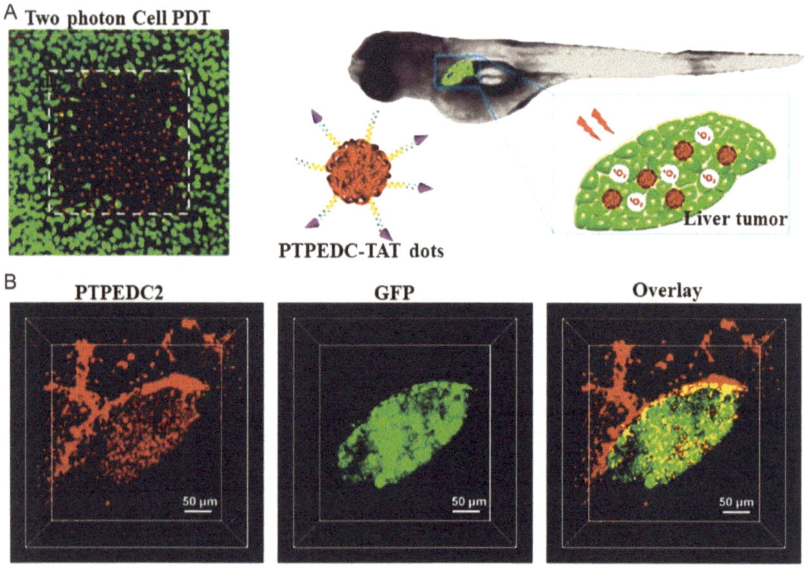

Fig. 7 (A) Represents the *in vivo* 2PE-PDT of **P23** dots in the liver tumor of zebrafish (right) and confocal live/dead cell imaging of HeLa cells incubated with **P23**-TAT dots subjected to 2PE-PDT (left). Emission: 575–620 nm (red, dead cells) and 500–545 nm (green, live cells). (B) 2 photon excited fluorescence (2PF) imaging of **P23** dots (**P23**, red) distributed in zebrafish liver tumor (GFP, green). Excitation: 820 nm; Emission: 635–675 nm (red) and 510–535 nm (green). *Adapted with the permission from Wang S, Wu W, Manghnani P, Xu S, Wang Y, Goh CC, Ng LG, Liu B. Polymerization-enhanced two-photon photosensitization for precise photodynamic therapy.* ACS Nano. *2019; 13:3095–3105.*

5.8-fold greater than that of its monomeric form and 2.7-fold greater for its donor-acceptor (D-A) even-odd effect. Importantly, the BT-TPAP NPs exhibit negligible dark toxicity because of its non-cationic characteristics which causes no depolarization of the mitochondrial membrane. Therefore, this strategy provides an efficient PS for imaging-guided PDT.[22]

5. Conclusion and outlook

AIEgens have emerged as a class of luminogenic materials with outstanding photophysical properties in the condensed state and explored for high-tech applications. Among the AIEgen's family, AIE active polymers exhibit unique properties of both the polymer and AIEgen such as structure versatility, high brightness, tunable emission, biocompatibility and processibility. This chapter presents the advances in AIE phenomenon and offers

AIE active polymers' design strategy with typical examples explored for biomolecules detection, *in vitro/in vivo* imaging, and image-guided therapy. The AIE active polymer can be designed by incorporation of AIEgens into the polymer backbone or side-chain via polycondensation, RAFT polymerization, and polycoupling reactions. AIE active polymers can also be constructed by insertion or grafting of AIEgens into the center or end of conventional polymers via post functionalization techniques. Through strategic design of AIE active polymers, turn on sensor can be fabricated for detection of biomacromolecules, bioanalyte, and diseases biomarker with high signal to noise ratio. Nevertheless, AIE active conjugated polymers exhibits tunable visible to near-infrared region (NIR) emission, high brightness and low background signal which helps in realization of wash free imaging and dynamic process monitoring with remarkable sensitivity and reliability. In contrast, AIE active non-conjugated polymers are non-covalently encapsulated or covalently conjugated with several prodrugs to form drug delivery carriers for image guided therapeutics applications. However, the AIE active polymers are still, emerging research discipline with magnanimous opportunities and requires consideration of following aspects to thrive further. First, the new polymerization techniques should be explored that leads to the development of wide variety of novel AIE active polymers. Second, narrow bandgap AIE active polymers with high absorption and brightness are especially needed for two photon, multiphoton and photoacoustic *in vivo* bioimaging and monitoring of dynamic cellular processes. Third, development of AIE active polymers with specific biorecognition units as the light up probes of bioanalytes and diseases biomarkers. Third, AIE active polymers with multimodal therapy such as PDT, chemotherapy and photothermal therapy (PTT) are needed for real time diagnosis and high quality therapeutics. Last but not the least, AIE active polymers with advance photophysical properties such as TICT, TADF, RTP should be explored for exciting research in materials and biological sciences.

References

1. Hong Y, Lam JWY, Tang BZ. Aggregation-induced emission: phenomenon mechanism and applications. *Chem Commun*. 2009;29:4332–4353.
2. Ding D, Li K, Liu B, Tang BZ. Bioprobes based on AIE fluorogens. *Acc Chem Res*. 2013;46:2441–2453.
3. Kwok RTK, Leung CWT, Lam JWY, Tang BZ. Biosensing by luminogens with aggregation-induced emission characteristics. *Chem Soc Rev*. 2015;44:4228–4238.

4. Liang J, Tang BZ, Liu B. Specific light-up bioprobes based on AIEgen conjugates. *Chem Soc Rev*. 2015;44:2798–2811.
5. Zhou L, Lv F, Liu L, Wang S. Water-soluble conjugated organic molecules as optical and electrochemical materials for interdisciplinary biological applications. *Acc Chem Res*. 2019;52:3211–3222.
6. Mei J, Leung NL, Kwok RT, Lam JW, Tang BZ. Aggregation-induced emission: Together we shine, united we soar! *Chem Rev*. 2015;115:11718–11940.
7. Wang H, Zhao E, Lam JW, Tang BZ. AIE luminogens: emission brightened by aggregation. *Mater Today*. 2015;18:365–377.
8. Gopikrishna P, Meher N, Iyer PK. Functional 1,8-naphthalimide AIE/AIEEgens: Recent advances and prospects. *ACS Appl Mater Interfaces*. 2018;10:12081–12111.
9. Zhang X, Wang K, Liu M, et al. Polymeric AIE-based nanoprobes for biomedical applications: Recent advances and perspectives. *Nanoscale*. 2015;7:11486–11508.
10. Hu R, Leung NLC, Tang BZ. AIE macromolecules: syntheses structures and functionalities. *Chem Soc Rev*. 2014;43:4494–4562.
11. Shen X, Shi Y, Peng B, et al. Fluorescent polymeric micelles with tetraphenylethylene moieties and their application for the selective detection of glucose. *Macromol Biosci*. 2012;12:1583–1590.
12. Wang K, Zhang X, Zhang X, et al. Fluorescent glycopolymer nanoparticles based on aggregation-induced emission dyes: preparation and bioimaging applications. *Macromol Chem Phys*. 2015;216:678–684.
13. Su X, Ma B, Hu J, et al. Dual-responsive doxorubicin-conjugated polymeric micelles with aggregation-induced emission active bioimaging and charge conversion for cancer therapy. *Bioconjug Chem*. 2018;29:4050–4061.
14. Liow SS, Dou Q, Kai D, et al. Long-term real-time in vivo drug release monitoring with AIE thermogelling polymer. *Small*. 2017;13:1603404.
15. Hirose A, Tanaka K, Yoshii R, Chujo Y. Film-type chemosensors based on boron diiminate polymers having oxidation-induced emission properties. *Polym Chem*. 2015; 6:5590–5595.
16. Hu R, Zhou T, Li B, et al. Selective viable cell discrimination by a conjugated polymer featuring aggregation-induced emission characteristic. *Biomaterials*. 2020;230: 119658.
17. Lv Y, Liu P, Ding H, et al. Conjugated polymer-based hybrid nanoparticles with two-photon excitation and near-infrared emission features for fluorescence bioimaging within the biological window. *ACS Appl Mater Interfaces*. 2015;7:20640–20648.
18. Zhang Z, Fang X, Liu Z, et al. Semiconducting polymer dots with dual-enhanced NIR-IIa fluorescence for through-skull mouse-brain imaging. *Angew Chem Int Ed*. 2020;59:3691–3698.
19. Wang Z, Wang C, Gan Q, Cao Y, Yuan H, Hua D. Donor–acceptor-type conjugated polymer-based multicolored drug carriers with tunable aggregation-induced emission behavior for self-illuminating cancer therapy. *ACS Appl Mater Interfaces*. 2019;11: 41853–41861.
20. Wang S, Wu W, Manghnani P, et al. Polymerization-enhanced two-photon photosensitization for precise photodynamic therapy. *ACS Nano*. 2019;13:3095–3105.
21. Zhou T, Hu R, Wang L, et al. An aie-active conjugated polymer with high ros-generation ability and biocompatibility for efficient photodynamic therapy of bacterial infections. *Angew Chem Int Ed*. 2020;59:9952–9956.
22. Liu S, Zhang H, Li Y, et al. Strategies to enhance the photosensitization: polymerization and the donor–acceptor even–odd effect. *Angew Chem Int Ed*. 2018;57:15189–15193.
23. Seo YH, Singh A, Cho H-J, et al. Rational design for enhancing inflammation-responsive in vivo chemiluminescence via nanophotonic energy relay to near-infrared AIE-active conjugated polymer. *Biomaterials*. 2016;84:111–118.

24. Liu S, Ou H, Li Y, et al. Planar and twisted molecular structure leads to the high brightness of semiconducting polymer nanoparticles for NIR-IIa fluorescence imaging. *J Am Chem Soc.* 2020;142:15146–15156.
25. Wang G, Zhang R, Xu C, et al. Fluorescence detection of DNA hybridization based on the aggregation-induced emission of a perylene-functionalized polymer. *ACS Appl Mater Interfaces.* 2014;6:11136–11141.
26. Liu L, Wu B, Yu P, Zhuo R-X, Huang S-W. Sub-20 nm nontoxic aggregation-induced emission micellar fluorescent light-up probe for highly specific and sensitive mitochondrial imaging of hydrogen sulfide. *Polym Chem.* 2015;6:5185–5189.
27. Wang Z, Chen S, Lam JWY, et al. Long-term fluorescent cellular tracing by the aggregates of AIE bioconjugates. *J Am Chem Soc.* 2013;135:8238–8245.
28. Wang T-T, Wei Q-C, Zhang Z-T, et al. AIE/FRET-based versatile PEG-Pep-TPE/DOX nanoparticles for cancer therapy and real-time drug release monitoring. *Biomater Sci.* 2020;8:118–124.
29. Huang J, He B, Zhang Z, et al. Aggregation-induced emission luminogens married to 2d black phosphorus nanosheets for highly efficient multimodal theranostics. *Adv Mater.* 2020;32:2003382.
30. Lin J, Zeng X, Xiao Y, et al. Novel near-infrared II aggregation-induced emission dots for in vivo bioimaging. *Chem Sci.* 2019;10:1219–1226.
31. Berry C, Brett M, Stevenson K, McMurray J, Norrie J. Nature and prognostic importance of abnormal glucose tolerance and diabetes in acute heart failure. *Heart.* 2008; 94:296–304.
32. Jovanovic L. The importance of postprandial glucose concentration: Lessons learned from diabetes and pregnancy. *Drug Dev Res.* 2006;67:591–594.
33. Yang L, Zhao M, Zhang R, et al. Synthesis and fluorescence study of a quaternized copolymer containing pyrene for DNA-hybridization detection. *ChemPhysChem.* 2012;13:4099–4104.
34. Wang G, Bobkov GV, Mikhailov SN, et al. Detection of RNA hybridization by pyrene-labeled probes. *Chembiochem.* 2009;10:1175–1185.
35. Szabó C. Hydrogen sulphide and its therapeutic potential. *Nat Rev Drug Discov.* 2007;6:917–935.
36. Yang G, Wu L, Jiang B, et al. H_2S as a physiologic vasorelaxant: hypertension in mice with deletion of cystathionine gamma-lyase. *Science.* 2008;322:587–590.
37. Kimura H. Hydrogen sulfide: Its production, release and functions. *Amino Acids.* 2011; 41:113–121.
38. Giorgio M, Trinei M, Migliaccio E, Pelicci PG. Hydrogen peroxide: A metabolic by-product or a common mediator of ageing signals? *Nat Rev Mol Cell Biol.* 2007;8: 722–728.
39. Gius D, Spitz DR. Redox signaling in cancer biology. *Antioxid Redox Signal.* 2006; 8:1249–1252.
40. Thubagere A, Reinhard BM. Nanoparticle-induced apoptosis propagates through hydrogen-peroxide-mediated bystander killing: Insights from a human intestinal epithelium in vitro model. *ACS Nano.* 2010;4:3611–3622.
41. Zehra N, Dutta D, Malik AH, Ghosh SS, Iyer PK. Fluorescence resonance energy transfer-based wash-free bacterial imaging and antibacterial application using a cationic conjugated polyelectrolyte. *ACS Appl Mater Interfaces.* 2018;10:27603–27611.
42. Hussain S, Lv F, Qi R, et al. Förster resonance energy transfer mediated rapid and synergistic discrimination of bacteria over fungi using a cationic conjugated glycopolymer. *ACS Appl Bio Mater.* 2020;3:20–28.
43. Loesche WJ, Lopatin DE, Stoll J, van Poperin N, Hujoel PP. Comparison of various detection methods for periodontopathic bacteria: can culture be considered the primary reference standard? *J Clin Microbiol.* 1992;30:418–426.

44. Ivnitski D, Abdel-Hamid I, Atanasov P, Wilkins E. Biosensors for detection of pathogenic bacteria. *Biosens Bioelectron.* 1999;14:599–624.
45. Lazcka O, Del Campo FJ, Muñoz FX. Pathogen detection: a perspective of traditional methods and biosensors. *Biosens Bioelectron.* 2007;22:1205–1217.
46. Li Y, Yu H, Qian Y, Hu J, Liu S. Amphiphilic star copolymer-based bimodal fluorogenic/magnetic resonance probes for concomitant bacteria detection and inhibition. *Adv Mater.* 2014;26:6734–6741.
47. Bovik AC. *Handbook of Image and Video Processing.* Academic Press; 2010.
48. Wu W, Bazan GC, Liu B. Conjugated-polymer-amplified sensing, imaging, and therapy. *Chem.* 2017;2:760–790.
49. Saadatpour Z, Rezaei A, Ebrahimnejad H, et al. Imaging techniques: new avenues in cancer gene and cell therapy. *Cancer Gene Ther.* 2017;24:1–5.
50. Lusic H, Grinstaff MW. X-ray-computed tomography contrast agents. *Chem Rev.* 2013;113:1641–1666.
51. Gallagher FA. An introduction to functional and molecular imaging with MRI. *Clin Radiol.* 2010;65:557–566.
52. Adil LR, Gopikrishna P, Iyer PK. Receptor-free detection of picric acid: a new structural approach for designing aggregation-induced emission probes. *ACS Appl Mater Interfaces.* 2018;10:27260–27268.
53. Wang Z, Wang C, Fang Y, Yuan H, Quan Y, Cheng Y. Color-tunable AIE-active conjugated polymer nanoparticles as drug carriers for self-indicating cancer therapy via intramolecular FRET mechanism. *Polym Chem.* 2018;9:3205–3214.
54. Arosio P, Albino M, Orsini F, et al. Multifunctional nanovectors based on polyamidoamine polymers for theranostic application. *J Nanosci Nanotechnol.* 2019;19: 5020–5026.
55. Feng H-T, Yuan Y-X, Xiong J-B, Zheng Y-S, Tang BZ. Macrocycles and cages based on tetraphenylethylene with aggregation-induced emission effect. *Chem Soc Rev.* 2018; 47:7452–7476.
56. Khatun MN, Tanwar AS, Meher N, Iyer PK. An unprecedented blueshifted naphthalimide AIEEgen for ultrasensitive detection of 4-nitroaniline in water via "receptor-free" IFE mechanism. *Chem Asian J.* 2019;14:4725–4731.
57. Shen J, Zhao L, Han G. Lanthanide-doped upconverting luminescent nanoparticle platforms for optical imaging-guided drug delivery and therapy. *Adv Drug Deliv Rev.* 2013; 65:744–755.
58. Langer R. Drug delivery and targeting. *Nature.* 1998;392:5–10.
59. Yoo J-W, Irvine DJ, Discher DE, Mitragotri S. Bio-inspired, bioengineered and biomimetic drug delivery carriers. *Nat Rev Drug Discov.* 2011;10:521–535.
60. Xu J, Yan B, Du X, et al. Acidity-triggered zwitterionic prodrug nano-carriers with AIE properties and amplification of oxidative stress for mitochondria-targeted cancer theranostics. *Polym Chem.* 2019;10:983–990.
61. Yuan Y, Xu S, Zhang C-J, Liu B. Light-responsive AIE nanoparticles with cytosolic drug release to overcome drug resistance in cancer cells. *Polym Chem.* 2016;7: 3530–3539.
62. Wang X, Xu K, Yao H, et al. Temperature-regulated aggregation-induced emissive self-healable hydrogels for controlled drug delivery. *Polym Chem.* 2018;9:5002–5013.
63. Wang H, Liu G, Gao H, Wang Y. A pH-responsive drug delivery system with an aggregation-induced emission feature for cell imaging and intracellular drug delivery. *Polym Chem.* 2015;6:4715–4718.
64. Wang H, Liu G, Dong S, Xiong J, Du Z, Cheng X. A pH-responsive AIE nanoprobe as a drug delivery system for bioimaging and cancer therapy. *J Mater Chem B.* 2015;3:7401–7407.

65. Wang X, Yang Y, Zhuang Y, et al. Fabrication of pH-responsive nanoparticles with an AIE feature for imaging intracellular drug delivery. *Biomacromolecules*. 2016;17: 2920–2929.
66. Dai Y-D, Sun X-Y, Sun W, et al. H_2O_2-responsive polymeric micelles with a benzil moiety for efficient DOX delivery and AIE imaging. *Org Biomol Chem*. 2019; 17:5570–5577.
67. Chen Y, Han H, Tong H, et al. Zwitterionic phosphorylcholine–TPE conjugate for pH-responsive drug delivery and AIE active imaging. *ACS Appl Mater Interfaces*. 2016;8:21185–21192.
68. Zhuang W, Xu Y, Li G, et al. Redox and pH dual-responsive polymeric micelles with aggregation-induced emission feature for cellular imaging and chemotherapy. *ACS Appl Mater Interfaces*. 2018;10:18489–18498.
69. Zhang C, Jin S, Li S, et al. Imaging intracellular anticancer drug delivery by self-assembly micelles with aggregation-induced emission (aie micelles). *ACS Appl Mater Interfaces*. 2014;6:5212–5220.
70. Zhuang W, Ma B, Liu G, Li G, Wang Y. TPE-conjugated biomimetic and biodegradable polymeric micelle for AIE active cell imaging and cancer therapy. *J Appl Polym Sci*. 2018;135:45651.
71. Wu P, Wang X, Wang Z, et al. Light-activatable prodrug and AIEgen copolymer nanoparticle for dual-drug monitoring and combination therapy. *ACS Appl Mater Interfaces*. 2019;11:18691–18700.
72. Dougherty TJ, Gomer CJ, Henderson BW, et al. Photodynamic therapy. *J Natl Cancer Inst*. 1998;90:889–905.
73. Wu W, Mao D, Xu S, et al. Polymerization-enhanced photosensitization. *Chem*. 2018;4:1937–1951.
74. Chilakamarthi U, Giribabu L. Photodynamic therapy: past, present and future. *Chem Rec*. 2017;17:775–802.
75. Yuan Y, Zhang C-J, Gao M, Zhang R, Tang BZ, Liu B. Specific light-up bioprobe with aggregation-induced emission and activatable photoactivity for the targeted and image-guided photodynamic ablation of cancer cells. *Angew Chem Int Ed*. 2015; 54:1780–1786.
76. Huang H, Yu B, Zhang P, et al. Highly charged ruthenium(ii) polypyridyl complexes as lysosome-localized photosensitizers for two-photon photodynamic therapy. *Angew Chem Int Ed*. 2015;54:14049–14052.
77. Lucky SS, Soo KC, Zhang Y. Nanoparticles in photodynamic therapy. *Chem Rev*. 2015;115:1990–2042.
78. Huang L, Li Z, Zhao Y, et al. Ultralow-power near infrared lamp light operable targeted organic nanoparticle photodynamic therapy. *J Am Chem Soc*. 2016;138:14586–14591.
79. Wu W, Shao X, Zhao J, Wu M. Controllable photodynamic therapy implemented by regulating singlet oxygen efficiency. *Adv Sci*. 2017;4:1700113.
80. Yuan Y, Kwok RTK, Tang BZ, Liu B. Smart probe for tracing cancer therapy: selective cancer cell detection, image-guided ablation, and prediction of therapeutic response in situ. *Small*. 2015;11:4682–4690.
81. Ethirajan M, Chen Y, Joshi P, Pandey RK. The role of porphyrin chemistry in tumor imaging and photodynamic therapy. *Chem Soc Rev*. 2011;40:340–362.
82. Guan Y, Lu H, Li W, et al. Near-infrared triggered upconversion polymeric nanoparticles based on aggregation-induced emission and mitochondria targeting for photodynamic cancer therapy. *ACS Appl Mater Interfaces*. 2017;9:26731–26739.
83. Xu W, Zhang Z, Kang M, et al. Making the best use of excited-state energy: multimodality theranostic systems based on second near-infrared (nir-ii) aggregation-induced emission luminogens (aiegens). *ACS Mater Lett*. 2020;2:1033–1040.

CHAPTER SEVEN

AIE-MOF materials for biological applications

Tahir Mehmood and J. Prakasha Reddy*
Department of Chemistry, School of Sciences, Indrashil University, Rajpur, Gujarat, India
*Corresponding author: e-mail address: j.prakashareddy@gmail.com

Contents

Abstract

Metal-Organic Frameworks (MOFs), coming under the realm of coordination chemistry, are unparalleled and the most studied among the group of porous materials. Structurally, these are well-defined three-dimensional crystalline products that can be tuned for various potential applications with a range of physico-chemical properties. More recently, aggregation-induced emission (AIE) and AIE of MOF material has attracted tremendous attention due to promising applications in biology. However, a chapter summarizing the work in AIE-MOFs materials has never been reported till date. A comprehensive review on the AIE and MOFs separately is beyond the reach of this chapter. Hence, we have summarized overview of recent developments in the syntheses and biological applications such as cell imaging, heparin detection, and drug delivery. In the end, conclusion, prospects and challenges in the arena of AIE-MOF materials are also highlighted.

1. Metal organic frameworks

In recent times, huge attention has been given to organic and metal-organic structures because of structural appeal and potential application especially in the chemistry-biology interphase.[1–6] A seminal work published by Alfred Werner in 1893 about the spatial arrangement of ligands

Progress in Molecular Biology and Translational Science, Volume 185
ISSN 1877-1173
https://doi.org/10.1016/bs.pmbts.2021.06.013

surrounding a metal ion and subsequently postulating the theory won him Nobel Prize in chemistry in 1913.[7] These complexes are usually referred in the literature as coordination complexes/polymers or Porous Coordination Polymers (PCPs).[8,9] Since past two decades, a new term coined by Yaghi referred as Metal-Organic Frameworks (MOFs), also referred as organic-inorganic hybrid material, is widely accepted/adopted among scientific community.[10] Due to rapid advances in the past several years, the field of MOFs has progressed to such an extent where a desired network structure can often be synthesized via coordination self-driven assembly by judiciously selected building blocks. Although, these networks were initially synthesized and studied from basic science point of view and for their fascinating, and esthetic architectural look but recent spike in the number of publications in this vital area of research show that these MOFs find applications including, and not limited to, sensing,[11] separation,[12] catalysis,[13,14] gas storage,[15] conductivity,[16] luminescence,[17] magnetism,[16] optics,[18] etc. More recently, burgeoning MOFs science has become a hot topic in the areas of biomedical application such as biosensing,[19] theranostics,[20] enzyme immobilization,[21] drug encapsulation, delivery, and controlled release.[22] By the combination of divergent choice of organic linkers (donors) and metal-containing nodes (acceptors), theoretically and practically as well, an infinite number of self-assembled metal complexes can be prepared. The advantage of the exceptional tunability of MOF material, both in synthesis as well as post synthetic modification,[23] along with knowledge gain in basics due to surge in number of publications allow these materials to remain one of the well-studied over other classes of materials. Since these coordination complexes are directional, and one can predict to some extent, the structural features with well-defined shapes and sizes (Fig. 1) could be constructed as is evident in the literature.[24]

To bring together more than one ligand molecule into an umbrella, i.e., construction of a single molecular entity by making use of two or more ligand molecules is a herculean task and remains a big challenge to organic chemists. Traditional organic synthesis suffers from complicated synthetic route, tedious and dismal purification. The synthetic route often involves large amount/number of solvents (not eco-friendly), final product often in low yield and seldom synthesized. However, technique/strategy of self-assembling the coordination complexes (formed by judicious combination of strapping together ligands and metal salts) remain a powerful tool and has been demonstrated which is evident from the literature through various robust and controlled shapes and sizes by individual research groups. Recently, Yaghi and co-workers reported a chronology of advances

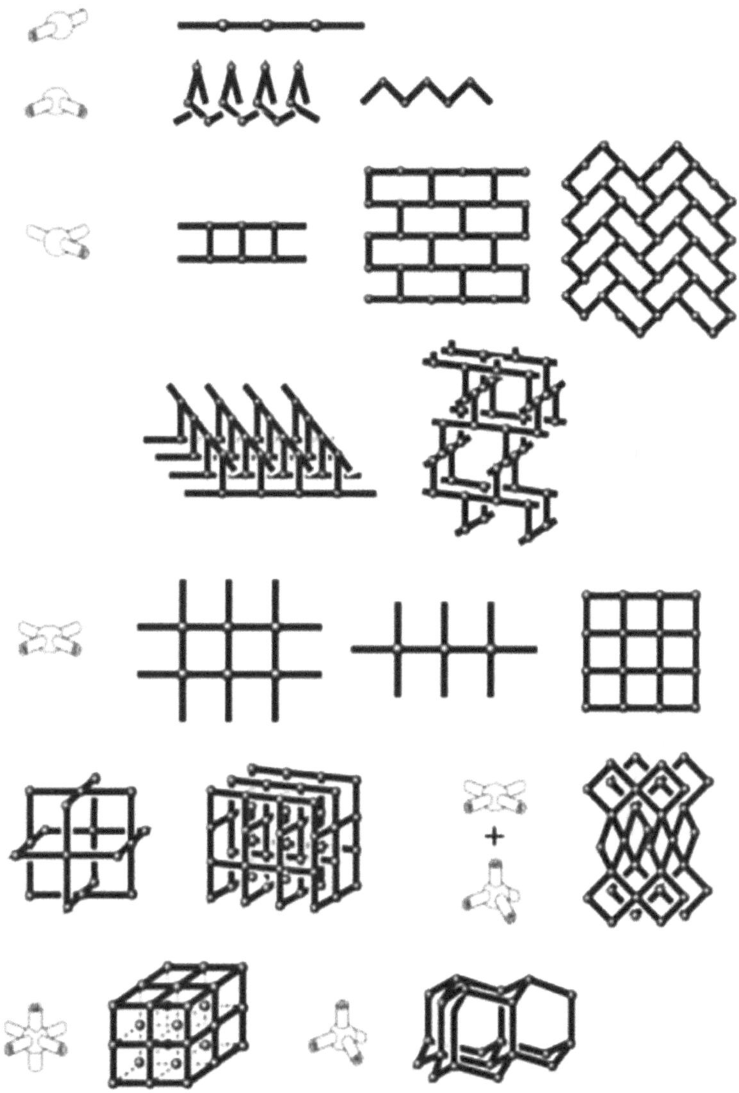

Fig. 1 Various structural frameworks that can be formed by using different donor and acceptor units. *Reprinted with permission from Kitagawa S, Kitaura R, Noro S. Functional Porous Coordination Polymers. Angew Chemie Int Ed. 2004;43:2334-2375. Copyright 2004 Wiley VCH.*

occurred starting from the discovery of metal complexes to MOFs till date in the realm of coordination chemistry (Fig. 2).[25] The advancement of knowledge aided in strategically controlling the extended network in three-dimension from the corresponding molecular components.

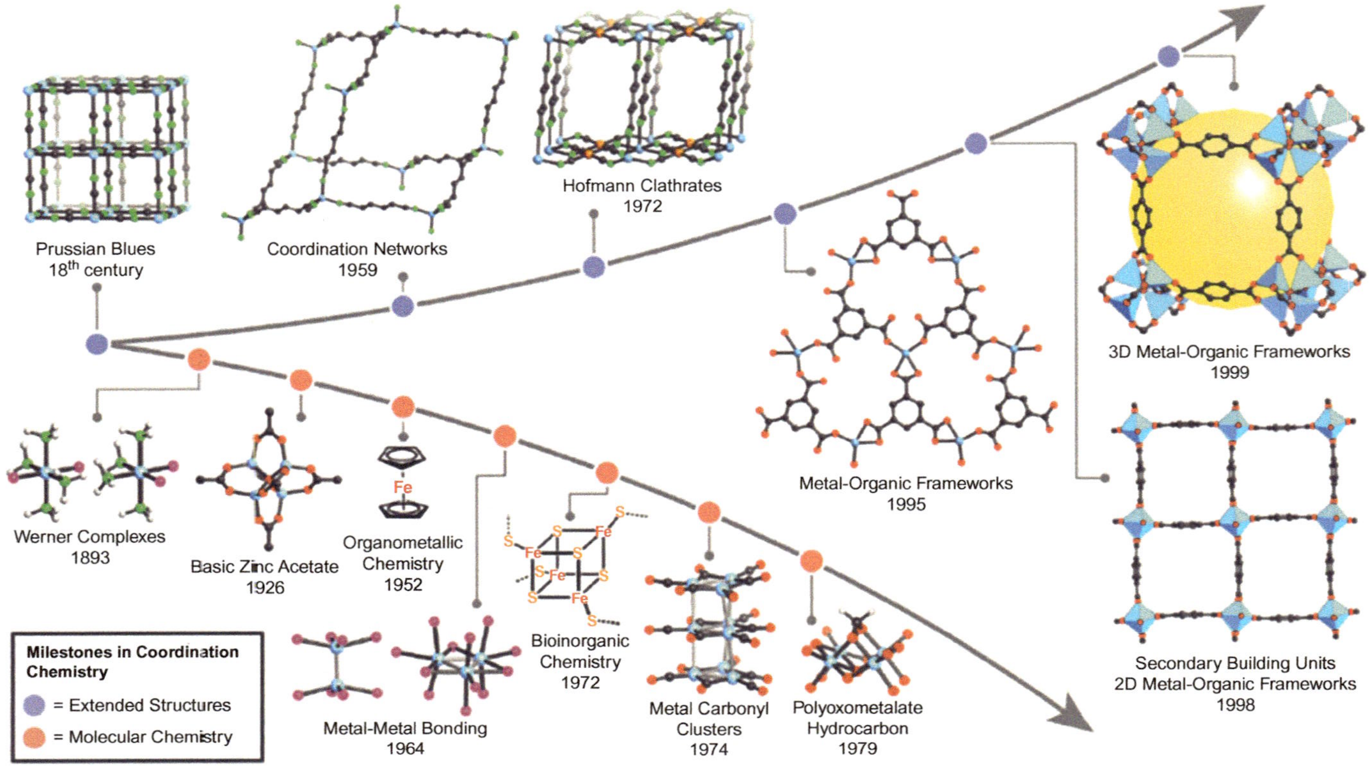

Fig. 2 Chronology of advances in coordination chemistry starting from observation of "Prussian Blue" compound and Werner theory of complexes to the metal-organic frameworks (MOFs). *Reprinted with permission from Diercks CS, Kalmutzki MJ, Diercks NJ, Yaghi OM. Conceptual advances from werner complexes to metal–organic frameworks.* ACS Cent Sci. *2018;4:1457–1464. Copyright 2018 American Chemical Society.*

As the choice of ligands vary infinitely, the position of coordinating group can also be varied and consequently results in strong, directional, and rigid framework with novel physico-chemical properties. There are numerous reports in the literature wherein using ligand with two carboxylate groups at 180° based (terephthalic acid) coordinating to metal leads to an extended network in three-dimension.[26] Similarly, ligands arranged at 120° having three carboxylate groups at symmetrical positions (1,3,5-benzenetricarboxylate) lead to the formation of honeycomb network as reported by numerous research groups.[27] In general, various carboxylic acids that differ in the position with different size and shapes of these molecules (as ligands) have been used in the construction of variety of MOFs. Though there can be many more possibilities of carboxylic acid ligands, some of the representative molecules have been shown in Table 1. The detailed study of these MOFs in terms of synthesis show that the principles on which Werner complexes were discovered, the same rules are applicable in the construction of MOF networks (and in general coordination complexes) although Werner complexes were built around the metal ion with counter ions.

2. Aggregation-induced emission of MOFs

In 2001, Tang and co-workers[48] first reported pioneering work on aggregation-induced emission (AIE) and has attracted wider attention from the scientific community because of potential applications.[49–52] Some representative examples of AIE luminogens (AIEgens) include, but not limited to, tetraphenylethylene (TPE), hexaphenylsilole, hexaphenylbenzene, derivatives of pyrrole, etc. These AIEgens, in general, emit no light in dilute solutions while in the solid (or condensed) state these chromophores emit light by molecular aggregation. For instance, TPE chromophore is well known that displays AIE character and has been introduced as a linker ligand in several MOF material. After a decade of the first report of AIE,[48] Dinca and co-workers,[53] in 2011, reported enhanced fluorescence quantum yield and lifetime of Zn (II) and Cd (II) carboxylate derivative of TPE linker-based AIE-MOF material (Fig. 3). Upon coordination of linker TPE ligand with metal, the rotation around the C—C bond is restricted (rotation freezes and π conjugation regained) resulting in brilliant emission and long-lived fluorescence. In the free state, which otherwise, π conjugation is lost due to free rotation of phenyl rings around the C—C bond resulting in non-emissive nature of material. A thorough experimental and theoretical investigation by the same research group concluded that the high emissive

Table 1 Various carboxylic acid ligands used in the construction of MOFs.

Sl. No.	Name	Structure	References
1	1,2-Benzenedicarboxylic acid	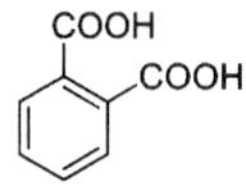	28
2	1,3-Benzenedicarboxylic acid	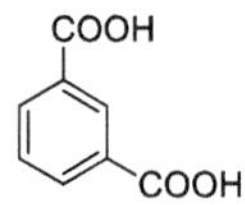	28,29
3	1,4-Benzenedicarboxylic acid	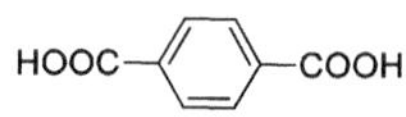	28–30
4	1,3,5-Benzenetricarboxylic acid	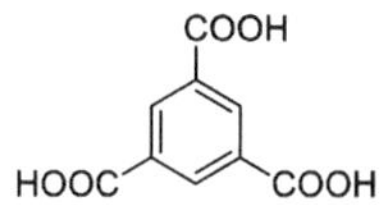	30,31
5	cis,cis-1,3,5-Cyclohexanetricarboxylic acid	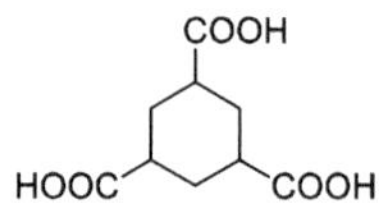	32
6	2,7-Pyrene dicarboxylic acid	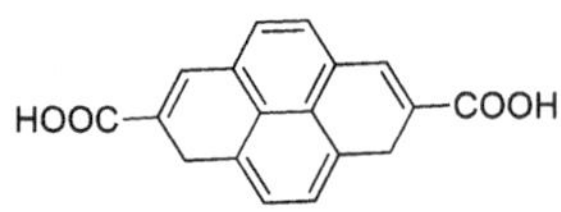	33
7	4,4′-Biphenyldicarboxylic acid	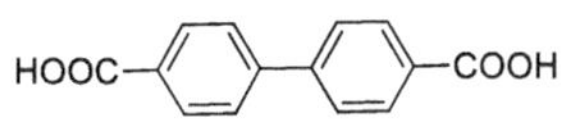	34
8	Trans-4,4′-Stilbenedicarboxylic acid	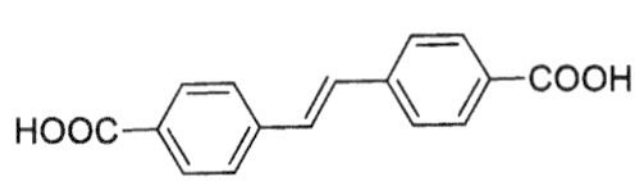	35
9	4,4′4″-Nitrilotrisbenzoic acid	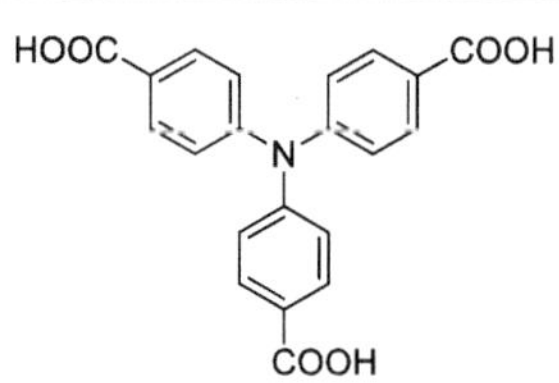	36

Table 1 Various carboxylic acid ligands used in the construction of MOFs.—cont'd

Sl. No.	Name	Structure	References
10	2,4,6-Tris(4-pyridyl)-1,3,5-triazine	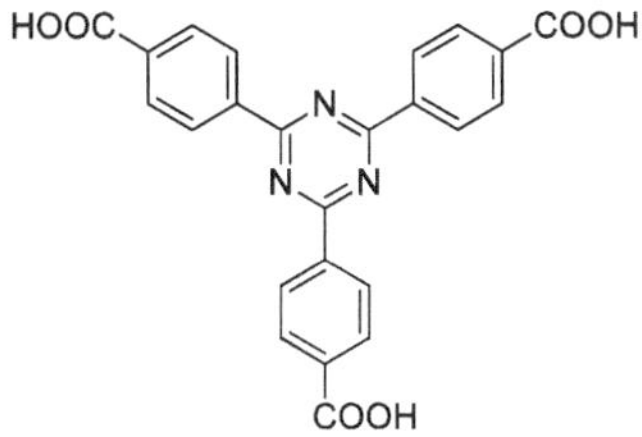	37
11	1,4-Naphthalenedicarboxylic acid	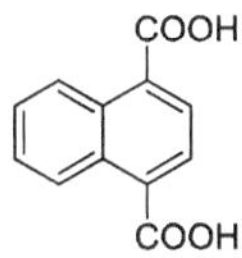	30
12	2,2′,3,3′-Oxydiphthalic acid	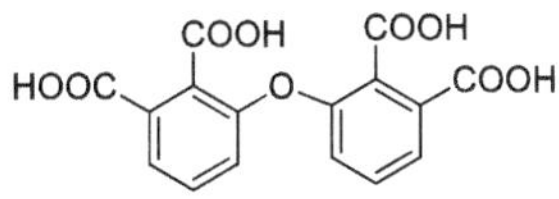	38
13	4,4′-Oxydiphthalic acid	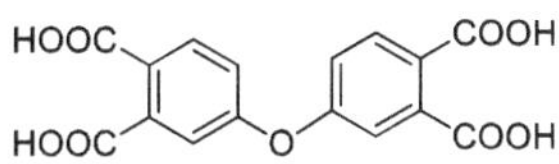	39
14	1,4-Phenylenediacetic acid	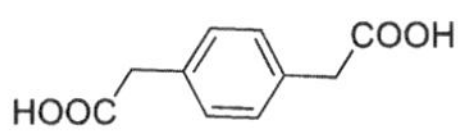	40
15	Nitrilotrisacetic acid	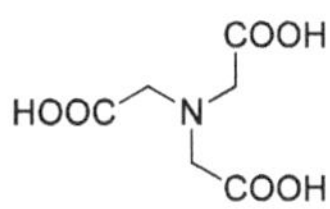	41
16	4,4′-Ethyne-1,2-diyldibenzoic acid		42
17	Fumaric acid	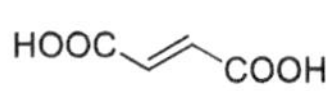	43
18	1,6-Hexanedioic acid	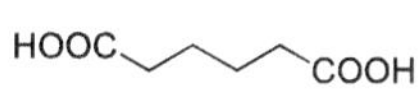	44
19	Oxalic acid	HOOC—COOH	45
20	3,5-Dinitrobenzoic acid	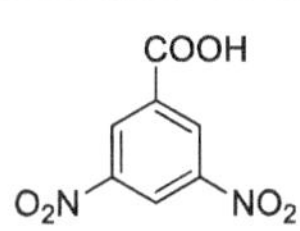	46
21	Pyridine-2,6-dicarboxylic acid	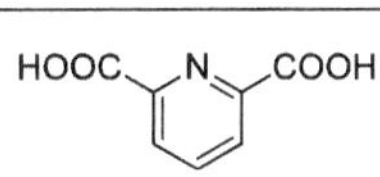	47

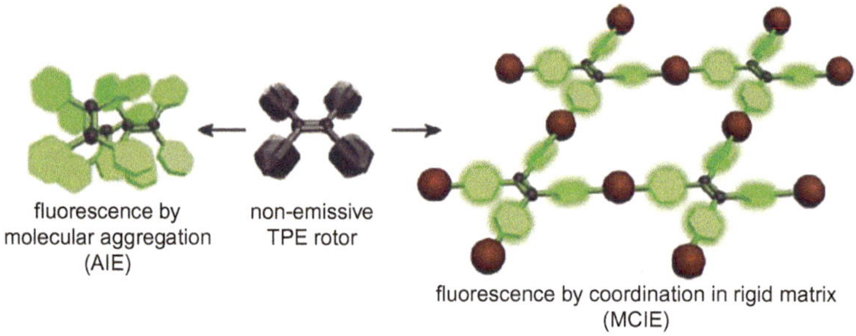

Fig. 3 Schematic representation of fluorescence by TPE molecule and Zn/Cd—TPE coordination complex. *Reprinted with permission from Shustova NB, Mccarthy BD, Dinca M. Turn-on fluorescence in tetraphenylethylene-based metal-organic frameworks: an alternative to aggregation-induced emission.* J Am Chem Soc. *2011;133:20126–20129.*

nature of material is due to high energy barrier for phenyl ring rotation. To the best of our knowledge, this is the first example where emissive linker ligand tethered by coordination bond is reported. In 2014, Zhou and co-workers[54] reported unusual photoluminescence properties of TPE-based Zr-MOF that can be tuned by rigidification of fluorescence linkers and structurally confining in MOF. In this work, they used Zr(IV) as metal node with modified TPE core unit as a linker and observed enhancement in the quantum yield (99.9%) and fluorescence lifetime. Due to these distinctive and interesting properties of the AIE of MOF material, special attention has been drawn more recently and created a new platform for research in the field of solid-state sciences/materials. This is because dynamic reversible architectural changes of MOFs to stimuli-response[55] (heat, pressure, light) and the structural retainment (strong crystallinity) endowed with designing and constructing of smart materials.

Metal-ligand bond is so strong that it provides robustness, restricted rotation, post-synthetic modification, etc., imparts intriguing, unique, and versatile properties to the material. These factors lead to the speculation that in addition to the applications of MOFs mentioned above, AIE-MOFs amenable to use in various research areas including biological applications. Because of directional bonding in coordination complexes, the structural features that lead to the formation of various supramolecular architectures can be predicted (at least to some extent) and subsequently material development for advanced applications. Since MOFs are porous in nature (permanent porosity), guest molecules will be absorbed and interact with chromophore/luminescent centers present in the proximity. The presence of guest (emissive) molecules in the pores results in change in the wavelength shifts, intensity, or sometimes even new emission. The increased emission lifetimes and quantum yield is also affected due to restricted moment of linker molecules (chromophore/lumophore) upon binding with metal that leads to a strong crystalline state which is not observed for the free complex in solution. In 2010, Rosseinsky and co-workers[56] reported a MOF with Indium as metal node and pyrene core linker based carboxylic acid (1,3,6,8-tetrakis(*p*-benzoic acid)pyrene) having permanent porosity (Fig. 4). Studies show that the MOF exhibits a good quantum efficiency of 6.7% and a fluorescence lifetime in the milliseconds range. Further structural analysis revealed that guest DMF molecules produces high degree of crystallinity and pyrene linkers are spatially well separated thereby minimizing the quenching effect and resulting in good emission. Interestingly, upon removal of guest molecules crystallinity of the MOF reduced (structural disorder or changes) and red-shifting of the emission was observed. On exposure of desolvated MOF material to DMF, a fully ordered (high crystalline) structure was regained with reversible fluorescence behavior. It is worth noting that the fluorescence lifetime of the free linker ligand was lower than In-MOF material in the solid state. On a similar note, Stang and co-workers,[57] in 2015, reported coordination-driven platinum (Pt) complex using multicomponent self-assembly strategy. The ligand used in this work is multipyridyl TPE core unit along with dicarboxylate ligand and Pt(II) as metal node to obtain AIE-active

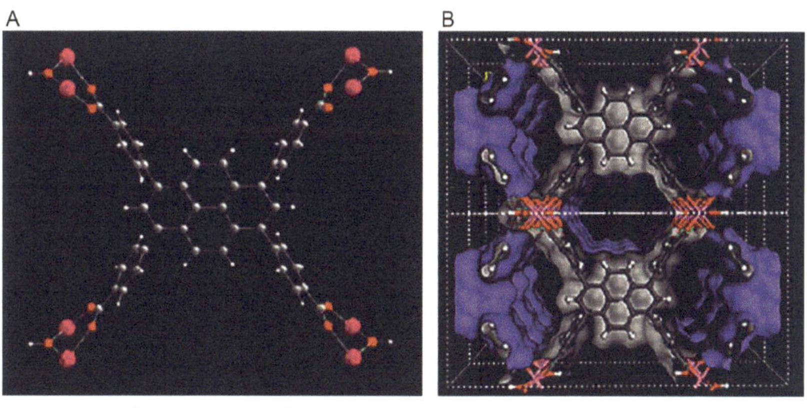

Fig. 4 (A) Coordination mode of TBAPy-In based complex, and (B) 3D noninterpenetrated structure of the material viewed along [010] direction. *Reprinted with permission from Stylianou KC, Heck R, Chong SY, et al. A guest-responsive fluorescent 3D microporous metal-organic framework derived from a long-lifetime pyrene core.* J Am Chem Soc. *2010;132:4119–4130.*

supramolecular complex which displayed 20-fold enhanced bright fluorescence compared to free ligand. These features, thus, advocate the advantages of AIE-MOF structures (having luminescent linkers) and allow the material to be tailored for a particular application.

3. AIE of MOF materials for biological applications

Since Tang's pioneering work on AIE, there are many reports of AIE and AIE-MOF material using modified/derivatives of TPE based core unit and other ligands which find biological applications such as selective probes, fluorescent labeling, high illumination regimes, cell imaging, diagnostics, etc.

3.1 Cell imaging

The cell being central dogma in all living organisms performs complex interplay of countless activities not only inside but outside as well. It provides a platform for several complicated metabolic reactions and/or pathways like catalysis (Ribozymes), regulate gene expression (Riboswitches), etc., involving DNA to RNA to proteins. Cell imaging, study of cells and analyzing signal dynamics (such as moment and migration of ions across cell membrane) has become increasingly important focus in biological research. This will provide a better understanding about how these dynamics affect all cellular decision-making activities in the biological system. However, it is utmost important to study cell without compromising on cytotoxicity and phototoxicity, a key in imaging technique. Inspired by the supra-amphiphiles[58–60] which act as potential biomaterials, Yang and co-workers[61] reported a TEP based supra-amphiphilic Pt(II) complex and further investigated the role of obtained biocompatible material in cell imaging. In this work, a new strategy based on directional bonding and symmetry principle interaction was implemented to prepare novel biomaterial by the combination of three 120° TPE-based core unit with three 120° di-Pt(II) metal node to form complex 6. This was followed by RAFT polymerization of 6 with NIPAM to form final material as shown in Fig. 5. Since the material is amphiphilic having both hydrophobic and hydrophilic units (hence biocompatible and water soluble) and AIE property, further studies were carried out to try and test the material as a potential candidate for cell imaging.

Assessment of cytotoxicity of the material is of paramount importance before any biomaterial could be applicable in biological system. The authors

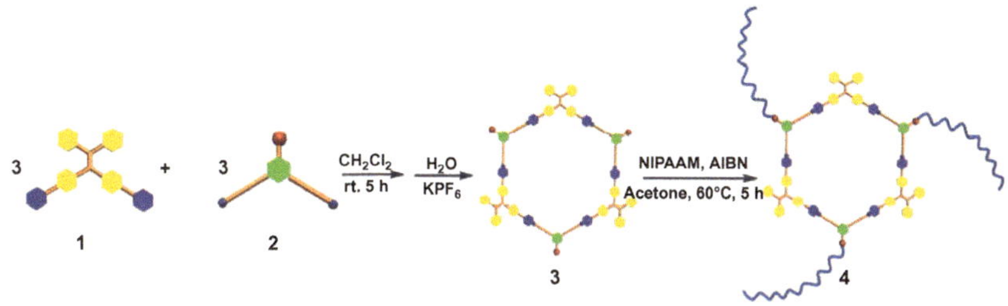

Fig. 5 Cartoon representation of a star type TPE-Pt based metal complex assembled from individual components. *Reprinted with permission from Zheng W, Yang G, Jiang S-T, et al. A tetraphenylethylene (TPE)-based supra-amphiphilic organoplatinum(ii) metallacycle and its self-assembly behaviour.* Mater Chem Front. *2017;1:1823–1828.*

carried out cytotoxicity measurement of the synthesized material on Bone Mesenchymal Stem Cells (BMSCs) using standard 3-(4,5-dimethylthiazol-2-yl)-2,5-diphenyltetrazoliumbromide (MTT) calorimetric assay. The results were found to be very encouraging as no change in cell viability was changed even upon increased concentration of the material. Further investigation, by confocal laser scanning microscopy (CLSM), proved that material was taken up by BMSCs. The green and blue-fluorescence signal arising from the material and 4′,6-diamidino-2-phenylindole (DAPI) respectively, observed in the same regions (Fig. 6) which indicated that the material was in the cell nucleus. In summary, material reported in this study has a good biocompatibility and find potential application in cell imaging.

3.2 Heparin detection

Heparin, a naturally occurring biomolecule with an average molecular weight of approx. 15 kDa and charge −70, is a linear, sulfated anionic polysaccharide used as an anticoagulant/antithrombotic (preventing blood clot formation) especially for the treatment of heart attacks and unstable angina. In 2006, heparin received worldwide attention, first, due to the death of three prematurely born infants due to mistaken overdose of the drug at Indianapolis hospital in USA. Second, in 2008, there was a major recall of the heparin drug due to contamination which took the lives of nearly 80 people in USA. Detection of heparin is not only important in the measurement of drug levels in the blood and clinical trials, for instance, bypass surgery, kidney dialysis, etc., but also to avoid mistaken overdose errors which could be fatal especially to pediatric patients. Although, there are reports of heparin sensing by potentiometric based selective electrodes

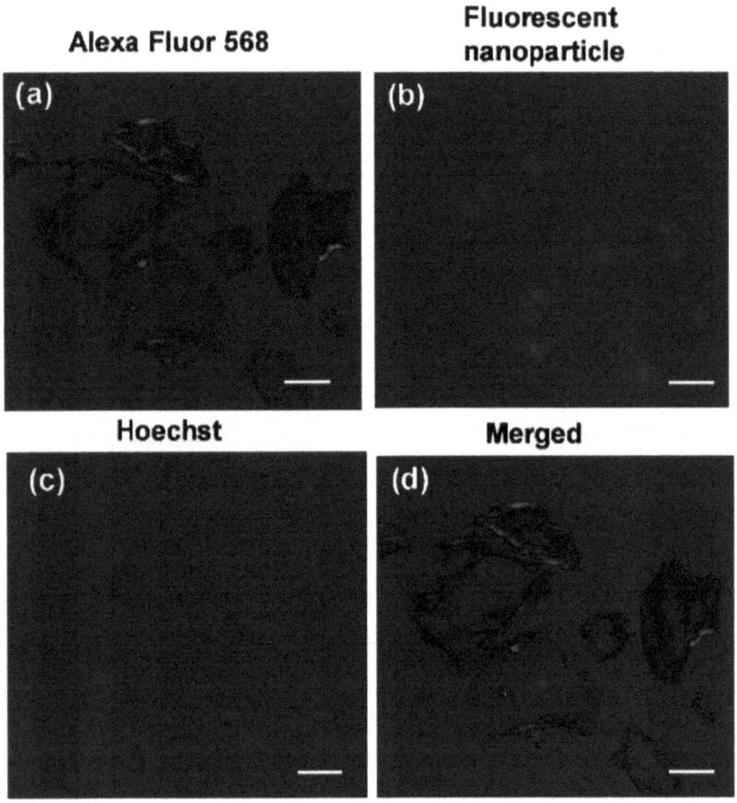

Fig. 6 Confocal laser scanning microscopy images of BMSCs cells incubated the fluorescent nanoparticles for material. (A) Alexa Fluor 568 red fluorescent signal attributed to BSSCs cells (B) green fluorescence signal from material in the nucleus, (C) Hoechst and (D) blue signal from DAPI stain, scale bar = 50 μm. *Reprinted with permission from Zheng W, Yang G, Jiang S-T, et al. A tetraphenylethylene (TPE)-based supra-amphiphilic organoplatinum(ii) metallacycle and its self-assembly behaviour.* Mater Chem Front. *2017;1:1823–1828.*

but has disadvantage due to high charge of heparin (~ -70) that results in potential drift and subsequently loss of sensor response.[62] However, AIE-MOF material has advantage over other sensing biomaterials. First, tuning the pore size is easy given the judicious choice of donor (ligands) and accepter (metal salts) molecules. Second, formation of coordination driven self-assembly is quite fast and hence ease of product synthesis is simple. Since MOFs are porous (presence of cavities and channels) in nature, incoming guest molecules can easily bind with host network through electrostatic interactions. Yang and co-workers,[63] in 2015, reported coordination complex using 120° TPE based dipyridine linker and 120° Pt(II) metal node (Fig. 7). Generally, this type of complexes possesses positive charges on

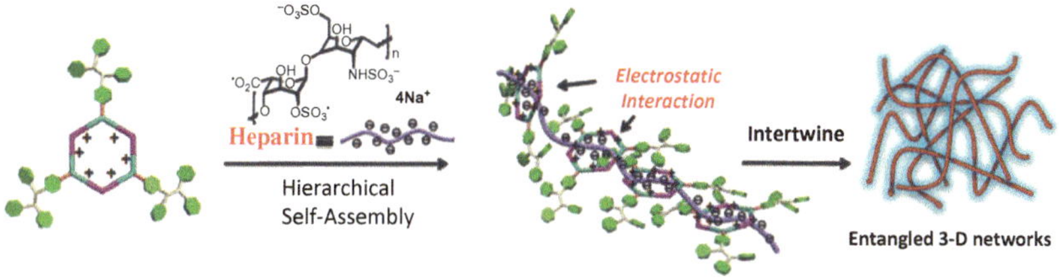

Fig. 7 Cartoon presentation of the possible binding site and aggregation of tris-TPE based Pt complex. *Reprinted with permission from Chen L, Ren Y, Wu N, et al. Hierarchical self-assembly of discrete organoplatinum (II) metallacycles with polysaccharide via electrostatic interactions and their application for heparin detection.* J Am Chem Soc. *2015;137:11725–11735.*

the core of the skeletal network and in this case, the complex has six positive charges distributed over core of the complex. Heparin was selected due to the presence of sulfate and carboxylate groups and hence it acquires high negative charge density. The positive charge present on metal complex skeleton interact with negative charge (sulfate and carboxylate groups of heparins) through electrostatic interactions. Upon the addition of heparin into acetone and water mixture of material, a gradually enhanced fluorescence was observed at 386 nm which was attributed to aggregation. Interestingly, when controlled experiment was performed, following the same procedure, luminescence was not observed for individual components. This study confirmed the presence of strong intermolecular interactions between metal complex and heparin.

Further support and better understanding in terms of the self-assembly of material was obtained from atomic force microscopy (AFM) studies. AFM images of dilute solution of material and heparin showed single bead-like chain with a height distribution ranging from 1.0 to 4.0 nm which was higher than individual components heparin and metal complex. Selectivity of material toward heparin was examined by recording fluorescence spectra with the addition of several other biomolecules such as sucrose (Suc), glucose (Glu), fructose (Fru), mannose (Man), galactose (Gal), lactose (Lac), maltose (Mal), arabinose (Ara), and serum albumins (bovine-BSA and human-HSA). Remarkably, indication for the good selectivity (out of a number of biomolecules) comes from increase in the emission intensity upon the addition of only heparin to complex while addition of other biomolecules showed no change in intensity of emission (Fig. 8). Additionally, chondroitin 4-sulfate and hyaluronic acid, which are structurally similar to heparin and common contaminants during clinical heparin procedures, were also chosen for selective examination. When added to complex, it was, however, observed that there was fluorescence emission but very much less in intensity compared

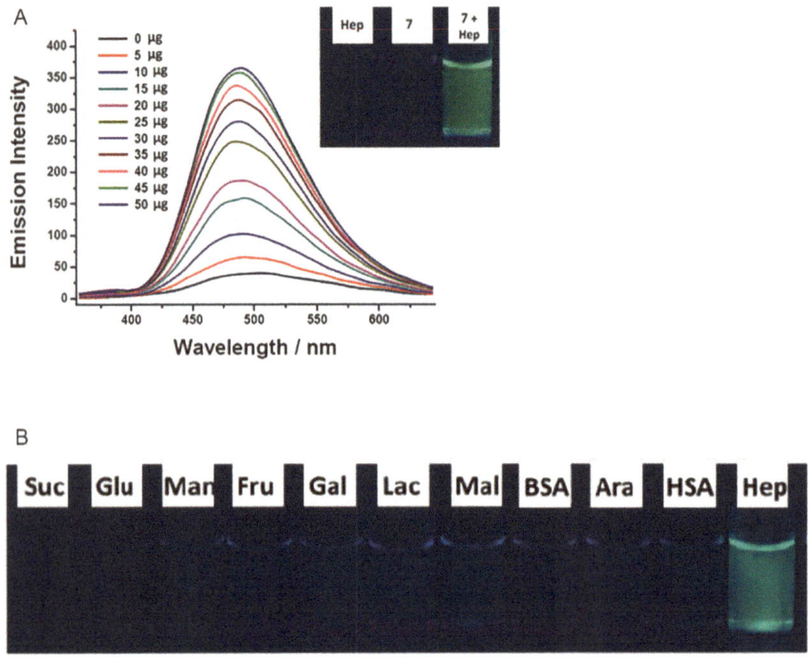

Fig. 8 (A) Observed emission spectral changes of Pt-trisTPE complex on heparin addition. Inset shows free heparin (left), Pt-trisTPE complex (middle) and their mixture (right) upon excitation at 365 nm at room temperature. (B) Images with other biomolecules such as Ara, arabinose; BSA, bovine serum albumin; Fru, fructose; Gal, galactose; Glu, glucose; Hep, heparin; HSA, human serum albumin; Lac, lactose; Man, mannose; Mal, maltose; Suc, sucrose. *Reprinted with permission from Chen L, Ren Y, Wu N, et al. Hierarchical self-sssembly of discrete organoplatinum (II) metallacycles with polysaccharide via electrostatic interactions and their application for heparin detection.* J Am Chem Soc. *2015;137:11725–11735.*

to heparin addition. These results proved that the complex was an excellent candidate for heparin selectivity.

3.3 Drug delivery

MOFs have been widely studied both in terms of drug encapsulation and delivery due to its porous nature. Drug molecules ranging from smaller size like ibuprofen, antibiotics, anticancer, antiviral molecules to bigger molecules such as nucleic acids and proteins have been incorporated in the porous networks.[64,65] Xie et al.,[66] in 2013, have reported Zirconium (Zr) based metal complex in which TPE core (H_4TCPE) unit was used as linker. The material was prepared by solvothermal process in presence of benzoic acid and HCl which found to have an average diameter of 500 nm having octahedron shapes. However, a well dispersed and a size less than 200 nm is required for the cellular uptake. Several smaller and larger sized nanoparticles were synthesized using different concentration of acetic acid as moderator at 90 °C. Dispersion of these nano-MOF particles in methanol resulted in color change from cyan to blue with UV lamp which was recorded by luminescent spectra at 360 nm. The result showed that the quantum yield increased from 0.061 to ~0.3 depending on the size of the MOF nano particles. Further, to examine the material, as potential behavior, curcumin (cur) was chosen as model anticancer drug. It is remarkable to note, in this study, the release of cur from cur-material in about 96 h which is 53.7% of cur release in acidic phosphate-buffer saline (PBS) solution and 40.4% in neutral PBS solution. Furthermore, to prove the biocompatibility of material, human cervix carcinoma cells, (HeLa), human lung cancer cells (A549) and human liver cancer cells (HepG2) were chosen, and further studies were carried out by a standard MTT assay. The observation of blue fluorescence by CLSM, show that the material nanoparticles are inside the cell rather on the surface. Cells were alive in all the three cell lines and no change in morphology of the cells was observed even after 48 h of incubation which shows no significant cytotoxicity. Anti-cancer activity was studied by incubating A549 cells with material nanoparticles, cur-material nanoparticles, and cur separately. CLSM images show that cur-material nanoparticles enter the cell (located in plasma) and quite a few of them even present in nucleus indicating that the material could deliver the cur into the cytoplasm (Fig. 9). The study proved that the complex is biocompatible and find potential application in drug loading and as a carrier.

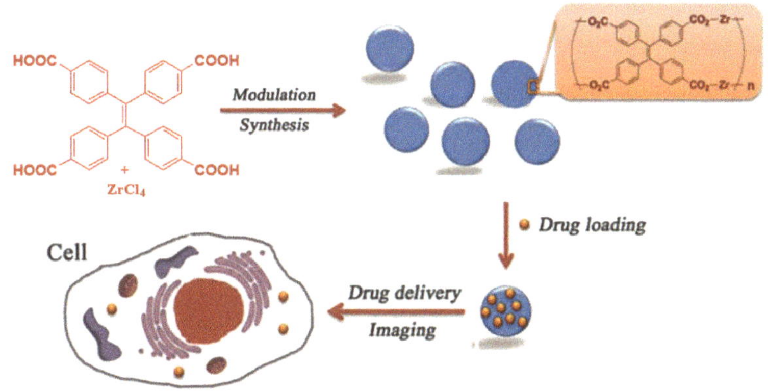

Fig. 9 Cartoon representation showing synthesis of Zr-TCPE MOF NCPs and its application in drug delivery and cell imaging. *Reprinted with permission from Wang L, Wang W, Xie Z. Tetraphenylethylene-based fluorescent coordination polymers for drug delivery.* J Mater Chem B. *2016;4:4263–4266.*

4. Summary and perspectives

Growing interest is given to AIE-MOF based material because of their appealing architectural features and various biological applications. Since last 5 years, significant advancement has been made with respect to synthesis and characterization of AIE-MOF functional solids. The combination of AIE-MOF based material is, however, complimentary than individual material/counterparts. It is noteworthy to mention that the material obtained will be more straightforward in application than individual AIE and MOF functional material. Harmony can be achieved between two different functionalities of AIE and MOFs counterparts that can lead to phenomenal application in biology and larger areas in science. The design and diversity in the judicious combination of ligands and metal salts, which may yield infinite number of novel materials, can be used precisely to target specific biological application. The extraordinary flexibility of AIE-MOFs offers surface functionalization/modification that could be of interest to bind biologically important molecules through electrostatic interactions. Both direct synthesis and post synthetic modification shows diversity in the AIE-MOFs derived material.

Several more factors need to be considered before the material can be applicable in biological system. The advances in the field of AIE-MOFs material are only still in their initial stages, endeavors should be devoted to understanding the mechanism of the aggregation induced crystallization in the solid state systematically. Additionally, it is noteworthy to mention that the AIE-MOF material with increased fluorescence quantum yield and longer lifetime should be paid special attention which result in enhanced property. Although a great number of AIE-MOF material have been reported since last 5 years, material with improved physico-chemical properties have been scarce. Also, the materials reported are active with respect to cell imaging, drug delivery, detection of biologically important molecules, etc., that mirrors the functionality, but origin of activity is poorly understood. In addition, more detailed information is required comprehensively at the structural level which is vital to understand all possible electrostatic interactions responsible for fluorescence of the material. In this direction, a perfect combination of both theoretical calculations and experimental studies will complement each other which can elucidate good understanding of structure-activity relationship with enhanced functionality of the material. Hence, attention should be given to the material that can demonstrate enhanced thermo-chemical stability that are strong and adequate to be used in various conditions especially at required pH in the biological system. Another most important issue to be addressed is water insolubility of AIE-MOF material for biological application due to which their usefulness is limited. Conceivably there is yet a long road ahead for high-rate commercial success of AIE-MOF material.

Acknowledgments

T.M. is thankful to SHODH (Scheme of Developing High quality research), Government of Gujarat, for financial support. We thank Dr. J. S. Yadav for his support and encouragement and Indrashil University for providing infrastructure.

References

1. Philp D, Stoddart JF. Self-assembly in natural and unnatural systems. *Angew Chem Int Ed Engl.* 1996;35:1154–1196.
2. Lehn BJ. Perspectives in supramolecular chemistry-from molecular recognition towards molecular information processing and self-organization. *Angew Chem Int Ed Engl.* 1990;29:1304–1319.
3. Thorat VH, Ingole TS, Vijayadas KN, et al. The ant-pro reverse-turn motif. Structural features and conformational characteristics. *Eur J Org Chem.* 2013;17:3529–3542.
4. Burslem GM, Kyle HF, Prabhakaran P, et al. Synthesis of highly functionalized oligobenzamide proteomimetic foldamers by late stage introduction of sensitive groups. *Org Biomol Chem.* 2016;14:3782–3786.
5. Deng H, Grunder S, Cordova KE, et al. Large-pore apertures in a series of metal-organic frameworks. *Science.* 2012;336:1018–1023.
6. Prakashareddy J, Pedireddi VR. Metal–organic hybrids of 1,3,5-Tris(4-pyridylsulfanylmethyl)-2,4,6-trimethyl-benzene with mercuric halides. *Eur J Inorg Chem.* 2007;2:1150–1158.
7. Werner A. Beitrag zur Konstitution anorganischer Verbindungen. *Z Anorg Chem.* 1893;3:267–330.
8. Janiak C. Engineering coordination polymers towards applications. *Dalton Trans.* 2003;14:2781–2804.
9. Kumar DK, Das A, Dastidar P. One-dimensional chains, two-dimensional corrugated sheets having a cross-linked helix in metal–organic frameworks: exploring hydrogen-bond capable backbones and ligating topologies in mixed ligand systems. *Cryst Growth Des.* 2006;6:1903–1909.
10. Yaghi OM, Li H. Hydrothermal synthesis of a metal-organic framework containing large rectangular channels. *J Am Chem Soc.* 1995;117:10401–10402.
11. Kreno LE, Leong K, Farha OK, Allendorf M, Van Duyne RP, Hupp JT. Metal–organic framework materials as chemical sensors. *Chem Rev.* 2012;112:1105–1125.
12. Li J-R, Sculley J, Zhou H-C. Metal–organic frameworks for separations. *Chem Rev.* 2012;112:869–932.
13. Yang D, Gates BC. Catalysis by metal organic frameworks: perspective and suggestions for future research. *ACS Catal.* 2019;9:1779–1798.
14. Huang Y, Wang Q, Liang J, et al. Soluble metal-nanoparticle-decorated porous coordination polymers for the homogenization of heterogeneous catalysis. *J Am Chem Soc.* 2016;138:10104–10107.

15. Li B, Wen H-M, Zhou W, Chen B. Porous metal–organic frameworks for gas storage and separation: what, how, and why? *J Phys Chem Lett.* 2014;5:3468–3479.
16. Alvaro M, Carbonell E, Ferrer B, Llabrøs FX. Semiconductor behavior of a metal-organic framework (MOF). *Chem A Eur J.* 2007;13:5106–5112.
17. Ren P, Liu M, Zhang J, et al. 1D, 2D and 3D luminescent zinc (II) coordination polymers assembled from varying flexible thioether ligands. *Dalton Trans.* 2008;320:4711–4713.
18. Zhang L, Qin Y, Li Z, et al. Topology analysis and nonlinear-optical-active properties of luminescent metal-organic framework materials based on zinc/lead isophthalates. *Inorg Chem.* 2008;47:8286–8293.
19. Wang H, Wang Y, Ding Y. Nanoscale Advances Development of biological metal-organic frameworks designed for biomedical applications: from bio-sensing/bio-imaging to disease treatment. *Nanoscale Adv.* 2020;2:3788–3797.
20. Yang J, Yang Y-W. Metal-organic framework-based cancer theranostic nanoplatforms. *View.* 2020;1:20.
21. Xia H, Li N, Zhong X, Jiang Y. Metal-organic frameworks: a potential platform for enzyme immobilization and related applications. *Front Bioeng Biotechnol.* 2020;8:695. https://www.frontiersin.org/article/10.3389/fbioe.2020.00695.
22. Cunha D, Ben Yahia M, Hall S, et al. Rationale of drug encapsulation and release from biocompatible porous metal–organic frameworks. *Chem Mater.* 2013;25:2767–2776.
23. Kalaj M, Cohen SM. Postsynthetic modification: an enabling technology for the advancement of metal–organic frameworks. *ACS Cent Sci.* 2020;6:1046–1057.
24. Kitagawa S, Kitaura R, Noro S. Functional porous coordination polymers. *Angew Chem Int Ed.* 2004;43:2334–2375.
25. Diercks CS, Kalmutzki MJ, Diercks NJ, Yaghi OM. Conceptual advances from Werner complexes to metal–organic frameworks. *ACS Cent Sci.* 2018;4:1457–1464.
26. Stock N, Biswas S. Synthesis of metal-organic frameworks (MOFs): routes to various MOF topologies, morphologies, and composites. *Chem Rev.* 2012;112:933–969.
27. Wee LH, Wiktor C, Turner S, et al. Copper benzene tricarboxylate metal–organic framework with wide permanent mesopores stabilized by Keggin polyoxometallate ions. *J Am Chem Soc.* 2012;134:10911–10919.
28. Du M, Jiang X-J, Zhao X-J. Molecular tectonics of mixed-ligand metal–organic frameworks: positional isomeric effect, metal-directed assembly, and structural diversification. *Inorg Chem.* 2007;46:3984–3995.
29. Yang E-C, Li J, Ding B, Liang Q-Q, Wang X-G, Zhao X-J. An eight-connected 3D lead(ii) metal–organic framework with octanuclear lead(ii) as a secondary building unit: synthesis, characterization and luminescent property. *CrstEngComm.* 2008;10:158–161.
30. Yang J, Li G-D, Cao J-J, Yue Q, Li G-H, Chen J-S. Structural variation from 1D to 3D: effects of ligands and solvents on the construction of lead(II)–organic coordination polymers. *Chem A Eur J.* 2007;13:3248–3261.
31. Chen W, Wang J-Y, Chen C, et al. Photoluminescent metal–organic polymer constructed from trimetallic clusters and mixed carboxylates. *Inorg Chem.* 2003;42:944–946.
32. Fang Q, Zhu G, Xue M, et al. Influence of organic bases on constructing 3D photoluminescent open metal–organic polymeric frameworks. *Dalton Trans.* 2004;14:2202–2207.
33. Allendorf MD, Bauer CA, Bhakta RK, Houk RJT. Luminescent metal–organic frameworks. *Chem Soc Rev.* 2009;38:1330–1352.
34. Wang X-L, Bi Y-F, Lin H-Y, Liu G-C. Three novel Cd(II) metal–organic frameworks constructed from mixed ligands of dipyrido[3,2-d:2′,3′-f]quinoxaline and benzenedicarboxylate: from a 1-D ribbon, 2-D layered network, to a 3-D architecture. *Cryst Growth Des.* 2007;7:1086–1091.
35. Bauer CA, Timofeeva TV, Settersten TB, et al. Influence of connectivity and porosity on ligand-based luminescence in zinc metal–organic frameworks. *J Am Chem Soc.* 2007;129:7136–7144.

36. Lee EY, Jang SY, Suh MP. Multifunctionality and crystal dynamics of a highly stable, porous metal–organic framework [Zn4O(NTB)2]. *J Am Chem Soc.* 2005;127: 6374–6381.
37. Li M-X, Miao Z-X, Shao M, Liang S-W, Zhu S-R. Metal-organic frameworks constructed from 2,4,6-tris(4-pyridyl)-1,3,5-triazine. *Inorg Chem.* 2008;47:4481–4489.
38. Zang S, Su Y, Li Y, Ni Z, Meng Q. Assemblies of a new flexible multicarboxylate ligand and d10 metal centers toward the construction of homochiral helical coordination polymers: structures, luminescence, and NLO-active properties. *Inorg Chem.* 2006;45: 174–180.
39. Chu Q, Liu G-X, Huang Y-Q, Wang X-F, Sun W-Y. Syntheses, structures, and optical properties of novel zinc(ii) complexes with multicarboxylate and N-donor ligands. *Dalton Trans.* 2007;4302–4311.
40. Soares-Santos PCR, Cunha-Silva L, Paz FAA, et al. Photoluminescent 3D lanthanide–organic frameworks with 2,5-pyridinedicarboxylic and 1,4-phenylenediacetic acids. *Cryst Growth Des.* 2008;8:2505–2516.
41. Chen Z, Zhao B, Zhang Y, Shi W, Cheng P. Construction and characterization of several new lanthanide–organic frameworks: from 2D lattice to 2D double-layer and to porous 3D net with interweaving triple-stranded helixes. *Cryst Growth Des.* 2008;8: 2291–2298.
42. Pham BTN, Lund LM, Song D. Novel luminescent metal-organic frameworks [Eu2L3(DMSO)2(MeOH)2]·2DMSO·3H2O and [Zn2L2(DMSO)2]·1.6H2O (L = 4,4′-ethyne-1,2-diyldibenzoate). *Inorg Chem.* 2008;47:6329–6335.
43. Zhu W-H, Wang Z-M, Gao S. Two 3D porous lanthanide–fumarate–oxalate frameworks exhibiting framework dynamics and luminescent change upon reversible de- and rehydration. *Inorg Chem.* 2007;46:1337–1342.
44. de Lill DT, Gunning NS, Cahill CL. Toward templated metal–organic frameworks: synthesis, structures, thermal properties, and luminescence of three novel lanthanide–adipate frameworks. *Inorg Chem.* 2005;44:258–266.
45. Cheng J-W, Zheng S-T, Yang G-Y. Diversity of crystal structure with different lanthanide ions involving in situ oxidation–hydrolysis reaction. *Dalton Trans.* 2007;36: 4059–4066.
46. Pedireddi VR, Varughese S. Solvent-dependent coordination polymers: cobalt complexes of 3,5-dinitrobenzoic acid and 3,5-dinitro-4-methylbenzoic acid with 4,4‘-bipyrdine. *Inorg Chem.* 2004;43:450–457.
47. Zhao B, Chen X-Y, Cheng P, Liao D-Z, Yan S-P, Jiang Z-H. Coordination polymers containing 1D channels as selective luminescent probes. *J Am Chem Soc.* 2004; 126:15394–15395.
48. Luo J, Xie Z, Lam JWY, et al. Aggregation-induced emission of 1-methyl-1, 2,3,4,5-pentaphenylsilole. *Chem Commun.* 2001;1740–1741.
49. Bhosale SV. Tetraphenylethene derivatives: a promising class of AIE luminogens—synthesis, properties, and applications. In: Bhosale RS, Aljabri M, La DD, Bhosale SV, Jones LA, Tang Y, Tang BZ, eds. *Principles and Applications of Aggregation-Induced Emission*; 2019:223–264. Springer International Publishing.
50. Cai X, Liu B. Aggregation-induced emission: recent advances in materials and biomedical applications. *Angew Chem Int Ed.* 2020;59:9868–9886.
51. Mei J, Leung NLC, Kwok RTK, Lam JWY, Tang BZ. Aggregation-induced emission: together we shine, united we soar! *Chem Rev.* 2015;115:11718–11940.
52. Kumar R, Yadav R, Kolhe MA, Bhosale RS, Narayan R. 8-Hydroxypyrene-1,3,6-trisulfonic acid trisodium salt (HPTS) based high fluorescent, pH stimuli waterborne polyurethane coatings. *Polymer (Guildf).* 2018;136:157–165.
53. Shustova NB, Mccarthy BD, Dinca M. Turn-on fluorescence in tetraphenylethylene-based metal-organic frameworks: an alternative to aggregation-induced emission. *J Am Chem Soc.* 2011;133:20126–20129.

54. Wei Z, Gu Z, Arvapally RK, et al. Rigidifying fluorescent linkers by metal–organic framework formation for fluorescence blue shift and quantum yield enhancement. *J Am Chem Soc.* 2014;136:8269–8276.
55. Gao X, Ge F, Zheng H. Improving the stability and visualizing the structural transformation of the stimuli-responsive metal–organic frameworks (MOFs). *Inorg Chem.* 2020;59:5093–5098.
56. Stylianou KC, Heck R, Chong SY, et al. A guest-responsive fluorescent 3D microporous metal-organic framework derived from a long-lifetime pyrene Core. *J Am Chem Soc.* 2010;132:4119–4130.
57. Yan X, Wang H, Hauke CE, et al. Tetraphenylethylene-based discrete organoplatinum (I) metallacycles: controllable structure and stoichiometry, aggregation-induced emission, and nitroaromatics sensing. *J Am Chem Soc.* 2015;137:15276–15286.
58. Wang C, Wang Z, Zhang X. Amphiphilic building blocks for self-assembly: from amphiphiles to supra-amphiphiles. *Acc Chem Res.* 2012;45:608–618.
59. Wang Y, Xu H, Zhang X. Tuning the amphiphilicity of building blocks: controlled self-assembly and disassembly for functional supramolecular materials. *Adv Mater.* 2009; 21:2849–2864.
60. Zhang X, Wang C. Supramolecular amphiphiles. *Chem Soc Rev.* 2011;40:94–101.
61. Zheng W, Yang G, Jiang S-T, et al. A tetraphenylethylene (TPE)-based supra-amphiphilic organoplatinum(ii) metallacycle and its self-assembly behaviour. *Mater Chem Front.* 2017;1:1823–1828.
62. Gemene KL, Meyerhoff ME. Reversible detection of heparin and other polyanions by pulsed chronopotentiometric polymer membrane electrode. *Anal Chem.* 2010;82: 1612–1615.
63. Chen L, Ren Y, Wu N, et al. Hierarchical self-assembly of discrete organoplatinum (I) metallacycles with polysaccharide via electrostatic interactions and their application for heparin detection. *J Am Chem Soc.* 2015;137:11725–11735.
64. Horcajada P, et al. Metal-organic frameworks as efficient materials for drug delivery. *Angew Chem Int Ed.* 2006;45:5974–5978.
65. Horcajada P, et al. Porous metal-organic-framework nanoscale carriers as a potential platform for drug delivery and imaging. *Nat Mater.* 2010;9:172–178.
66. Wang L, Wang W, Xie Z. Tetraphenylethylene-based fluorescent coordination polymers for drug delivery. *J Mater Chem B.* 2016;4:4263–4266.

CHAPTER EIGHT

Patented AIE materials for biomedical applications

Sopan M. Wagalgave[a,b], Shailesh S. Birajdar[a,b], Jotiram N. Malegaonkar[a,b], and Sidhanath Vishwanath Bhosale[a,b,*]

[a]Polymers and Functional Materials Division, CSIR-Indian Institute of Chemical Technology, Hyderabad, Telangana, India

[b]Academy of Scientific and Innovative Research (AcSIR), Ghaziabad, Uttar Pradesh, India

*Corresponding author: e-mail address: bhosale@iict.res.in

Contents

Abstract

In recent years aggregation induced emission (AIE) concept has attracted researcher's interest worldwide. Several organic building blocks are developed as AIE materials. This chapter discusses the patented AIE material and their utilization related in biological, medicinal and biotechnology fields. It is demonstrated that AIE chromophores such as tetraphenylethylene (TPE) as well as other AIE building blocks became important fluorescent emissive bioactive materials. Such emissive materials are widely employed as bioprobes for the detection of mitochondria, cellular imaging and tracking, protein carrier detection of S-phase DNA, detection of D-glucose, visualization of cancer treatment, drug screening, image-guided therapy, bacterial imaging, photodynamic therapy and drug screening. Such AIE materials upon imaging in cellular environment displays significant enhancement of fluorescence emission. Such patented AIE chromophores

Progress in Molecular Biology and Translational Science, Volume 185
ISSN 1877-1173
https://doi.org/10.1016/bs.pmbts.2021.07.003

has a great potential for bioimaging and biomedical applications. In this chapter we compile some patented representative examples to explore their bioimaging/medicinal imaging applications since lot of new inventions are reported every day.

1. Introduction

In the evolutionary process of life on the earth planet, nature developed living organisms exhibiting luminescence.[1] The term luminescence was derived from the Latin word *lumen* (light).[2] Luminescence is a spontaneous process and emits radiation upon excitation of electronic species. Fluorescence, phosphorescence, chemiluminescence, electroluminescence, radioluminescence, cathodoluminescnce and bioluminescence process comes under the term luminescence.[3] These processes are important in natural systems. Herein, we will deal with the fluorescence emission characteristics. The fluorescence observed from natural systems was first reported by the Spanish of new medical herbs.4 Spanish physicists and botanists Nicolás Monardes in 1565 reported a peculiar blue color but he did not know the color was blue fluorescence.4 Furthermore, Franciscan Friar Bernardino de Sahagún (1499–1590) also cited this wood of the species Eysenhardtia polystachya as "*coatli*."[5,6] It was demonstrated that Metaline was responsible for blue fluorescence with an emission of about 466 nm. But recently it was proven that the Metaline is resulted from the oxidation of flavonoids.[7] Based on these observations up to 1852 using non-specific terms as dispersion the phenomenon of fluorescence was vaguely defined. David Brewster in 1833 detected the fluorescence from the natural chromophore chlorophyll and the fluorescence process was described by him.[8] He demonstrated that when a sun light beam was passed through an alcohol solutions of leaves, an emitting red beam could be observed.[9] He also reported re-absorption of fluorescence. In 1845, Herschel demonstrated fluorescence characteristics of Quinine, isolated from the bark of *Cinchona* tree, solution and are termed as "epipolic dispersion."[10,11] Stokes, in 19th century reported emission properties of quinine using prism to disperse light.[12] Literature search revealed that the compounds present in biological systems has been related to the fluorescence process. It was well documented that leaves, flowers, fruits, avians, bees, beetles, butterflies, cockroaches, dragonflies, millipedes, spiders, scorpions and sea organisms exhibit fluorescence emitting properties.[13] In these natural systems, fluorescent natural products might play an important role during evolutionary process and such chemicals open new research directions in recent years. Therefore, such fluorescent probes have been utilize for exploring several biological aspects of cellular biology. They can be stimuli responsive and display changes in emission properties upon changing temperature, pH, ionic strength, viscosity, hydrophobicity, chelation effects with biomacromolecules such as DNA, lipids and specific proteins.[13] Such fluorophores are chosen to target subcellular compartments by means of chemical labeling.

1.1 Mechanism of fluorescence spectroscopy

When short-wavelength electromagnetic radiations are absorbed by the chromophores and then remitted at longer wavelength and the phenomenon is called as fluorescence. The fluorescence phenomenon was demonstrated by Polish physicists, Aleksander Jablonski and is named as Jablonski energy diagram (Fig. 1).[14] The description of the light absorption and emission with radiative and no-radiative transmission was given by him.

In first step, the light is absorbed by the chromophore and an electron is excited to higher singlet energy state. From higher singlet state via non-radiative transition led to reach it at singlet excited (S1) state. Thus internal conversion to S1 state make sit available to go to ground state or transfer to higher triplet state via intersystem crossing process. When electron from its S1 state reaches to singlet ground (S0) state, the phenomenon is called as fluorescence.[15]

In recent decades, the development of novel chromophores with fluorescence emission characteristic are an attractive material due to their promising applications in various fields such as sensors, chemosensors, biosensors, organic filed effect transistors, organic light emitting diodes (OLED), etc.1–4 Chromophores exhibit good absorption and emission properties in its highly dilute solutions. These chromophores in its higher

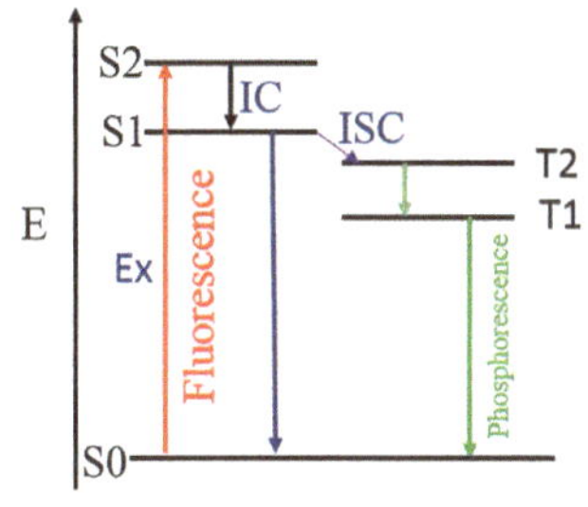

Fig. 1 A typical Jablonski energy diagram (S-singlet state, T-triplet state, IC-internal conversion and ISC-intersystem crossing).

concentration as well as display fluorescence quenching phenomenon. The fluorescence quenching process occurs via static or dynamic quenching.[20] Before excitation, an emissive chromophore and quencher are creating non-fluorescent complex led to quench the fluorescence and called as static quenching. The quenching of fluorescence occurs between two light-sensitive chromophores such as donor-acceptor through exchange of energy upon excitation. Herein, acceptor may quench the energy of the donor molecules. The process is explained with different energy transfer processes such as Dexter energy transfer, Förseter energy transfer and exciplex. The fluorescence quenching occurs via aggregation of the chromophores are called as Aggregation Caused Quenching (ACQ). In aggregated state, chromophore undergoes strong intermolecular π-π stacking interactions between the neighboring fluorophores. The excited states of such molecular aggregates relax back via non-radiative decay process leading in to the emission quenching of the chromophores.[21] ACQ materials display lower quantum yield in its aggregated state. Therefore, such ACQ material has restricted real-world applications. To overcome ACQ effect and develop chromophores with exhibiting properties such as highly emissive in the solid state, in aggregation state and in solid state is an attractive topic of researchers.

1.2 Aggregation induced emission (AIE) properties

Very recently Prof. Würthner compiled the historical perspective of aggregation induced emission (AIE) phenomenon.[22] He systematically demonstrated how AIE molecular architectures were developed from the 20th century and studied their fluorescence enhancement properties by various researchers. In 2001, for the first time AIE nomenclature of fluorescence enhancement phenomenon was coined by Prof. Tang et al. They demonstrated that 1-methyl-1,2,-3,4,5-pentaphenylsilole (**1**) solution in ethanol display an increase in fluorescence emission with the addition of water (Fig. 2).[23] The AIE phenomenon is reverse of the ACQ effect. In AIE, the chromophores are highly emissive at its solid state as well as an aggregated state in solution. Very recent years, development of AIE materials with such interesting emissive behavior in aggregation and solid states are widely employed for a range of applications including sensors, chemosensors, biosensors, bioimaging, cancer treatment, photodynamic therapy and in materials filed such as OFETs and OLEDs.[24]

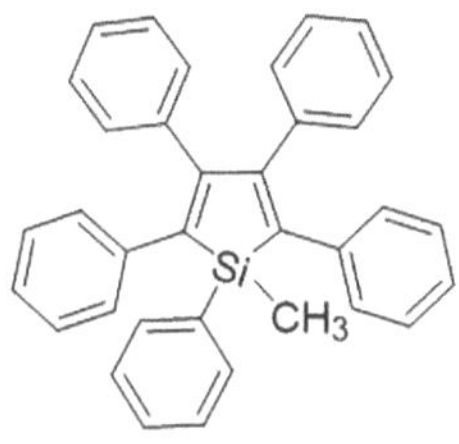

1

Fig. 2 Structure of 1-methyl-1,2,-3,4,5-pentaphenylsilole.

1.3 Aggregation induced emission mechanism

For the bioimaging and materials technology applications it is necessary to understand the photophysical characteristics of AIE molecular architecture. In order to take benefits of the novel AIE materials, the mechanism of AIE phenomenon has been extensively investigated. Tetraphenylethylene (TPE) was utilized to understand the AIE phenomenon. The excited AIE chromophores in solution state decay through non-radiative decay process, whereas, in aggregate state via radiative process. As demonstrated in Fig. 3, four phenyl rings are attached to an olefin subunit via single C—C bonds. In dilute solutions, when TPE chromophore is excited by the irradiation the aryl ring systems attached to the olefinic double bond are rotated with great freedom and the photonic energy is transformed to the thermal energy, which leads to radiationless relaxation.[26] Whereas, TPE chromophore under aggregated state shows restricted rotation of the phenyl subunits. Herein, intermolecular π-π staking is not observed due to twisted molecular confirmation. This led to the radiative decay for the energy of the excited molecular architecture. The radiative decay of the energy resulted into the enhanced emission of TPE via the "restriction of intramolecular motions (RIM)."[25] It is well demonstrated that the RIM processes includes both restriction of intramolecular rotation (RIR) and restriction of intramolecular vibrations (RIV) phenomenon.[25]

The AIE phenomenon opens up new area of research for the design, synthesis and development of AIE materials with diverse range. The AIEgens have been employed for different applications such as biosensing, chemical sensing, OLED, organic field effect transistor (OFET), solar cells and stimuli responsive technological applications.[24–26] To achieve high selectivity, sensitivity AIE chromophores can be utilized in concentrated situation. In recent years, the increasing AIE research globally resulted into considerable number of technical and scientific papers along with patents.

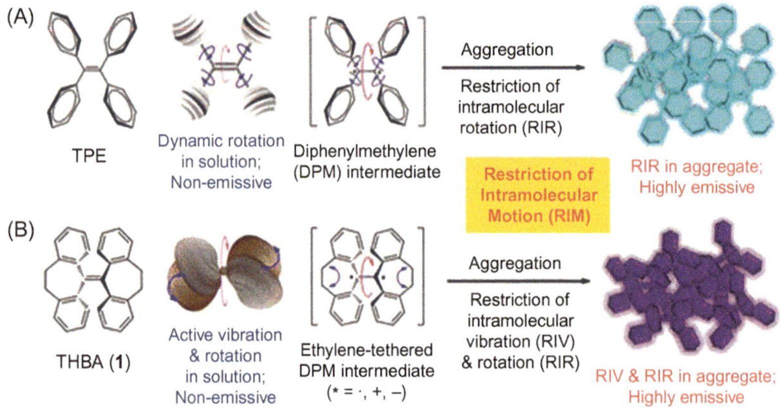

Fig. 3 The mechanism for AIE properties via (A) the restriction to the intramolecular rotations (RIR); (B) the restriction to the intramolecular vibration (RIV). *Adapted and modified with permission from Mei J, Hong YN, Lam JWY, Qin AJ, Tang YH, Tang BZ. Aggregation-induced emission: the whole is more brilliant than the parts.* Adv Mater. *2014; 26: 5429–5479. Copyright 2014 Wiley-VCH.*

2. Biomedical applications

In biology, bioimaging is a technique and process in which visual detection of interior of the cells, tissues or body for biological analysis and medical intervention is performed. Bioimaging is utilized in medical purpose so termed as biomedical imaging. It is demonstrated that biomedical imaging technology plays key role in modern medicine. To save human life, people have the need of advanced technologies based on biomedical imaging. Therefore, in recent years, biological imaging research make a pace for saving lives.[27] The literature and market search revealed that there are several imaging technologies such as magnetic resonance imaging (MRI), optoacoustic imaging, ultrasound imaging, photoacoustic imaging, X-ray radiography, positron emission tomography, fluorescence imaging and bioluminescence imaging are available.[27,28] Among these technologies, fluorescent imaging technique is an attractive and led to more and more research work in recent years.[27,28] Fluorescent biomedical imaging technique exhibits several merits such as variety of fluorescent molecular building blocks, low cost, intense output signal and wide applicability. It is already noted that the conventional fluorescent molecular architecture display ACQ effect,[21] which in turn prevents them from being their large uses in biological imaging. Therefore, AIEgens in the aggregated and solid state shows strong fluorescence emission and make them promising chromophores for bioimaging, i.e., biomedical applications.[24] It was observed that the AIEgens utilized in biomedical applications show superior properties, e.g., considerable absorptivity, high brightness, free of random blinking, strong photobleaching resistance and excellent biocompatibility compared to conventional organic chromophores.[24–26] These merits exhibited by AIEgens have enabled them to use in biological systems for fluorescent imaging of cellular, subcellular and tissue level.[24,29] AIEgens such as organic NPs, polymeric NPs, AIE dots, silica NPs are an attractive candidates for dual or multiple functions in biological/medical world of imaging.[29–32]

Herein, we summarize the patents filed by researchers for the protection of the intellectual property of their invention to have its commercial rights. In the recent years, AIEgens and their biomedical application patent filing and commercialization has a remarkable growth. AIE based biomedical imaging technology will bloom with new companies and startups looking for profit from these technologies.

For the benefit of the researchers, readers and industrialist, in this chapter we mainly focused to compiled the patented AIEgens employed for wide biological and biomedical applications.

2.1 Mitochondria imaging

In eukaryotic cells, a double membrane-bound organelle named as mitochondrion is present.[33] Mitochondrion is involved in the production of adenine triphosphate utilizing oxygen.[34] It has been the major energy source of the living cell. Apart from this work function, mitochondria are also involves in several biological processes such as signaling, cellular differentiation, control of the cell cycle, growth and cell death.[35,36] Mitochondria are participated in the events of reactive oxygen species (ROS), which can damage the membrane, which in turn is responsible for cell signaling and death.[37,38] The cell apoptosis induced several neurodegenerative diseases, e.g., Parkinson's disease, Alzheimer's disease and neurodegenerative disorder, etc.[39] In order to avoid the oxidative damage of the cell the mitochondrial glutathione (GSH) plays an important role as an antioxidant within the cell. The concentration of GSH is sensitive to oxidative stress of the mitochondrial.[40] Therefore, it is of great importance to design and develop fluorescent probes to detect the mitochondrial biothiols in living cells.[41,42] It has been proven fluorescent probes are an effective tool for the detection of the mitochondrial GSH and are less satisfactory in terms of photostability.[43]

In the past few years, to overcome the limitations of ACQ materials researcher developed new AIE molecular architectures to detect mitochondrial in living cells and patented (Table 1).[44]

In this patent, the investigators claimed the synthesized AIE derivatives such as TPE-TPP (**2**), TPE-MitoR (**3**) and TPE-IQ (**4**) are employed as fluorescent chromophores for mitochondrial imaging as apoptosis inducers (Table 1). These TPE derivatives are specifically detect mitochondria with

Table 1 Mitochondrial imaging.

Sr. No.	Representative example of AIEgens	Applications	References
1	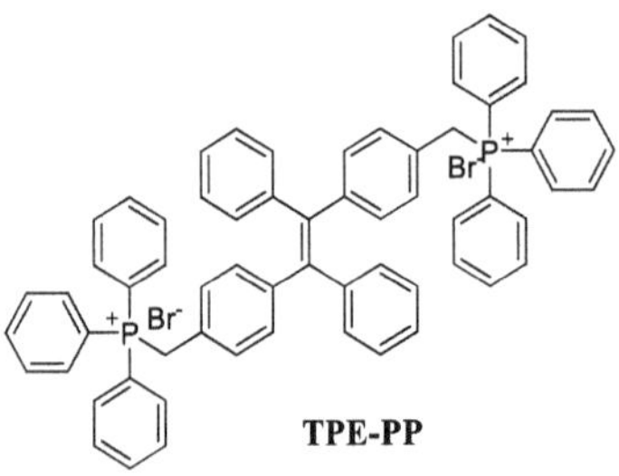 **TPE-PP** **2** 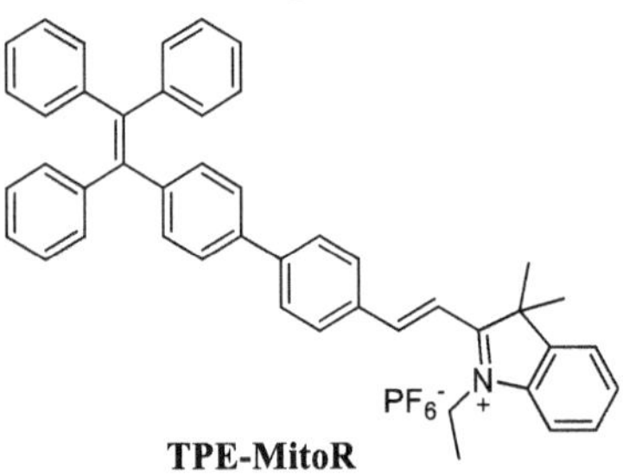 **TPE-MitoR** **3** 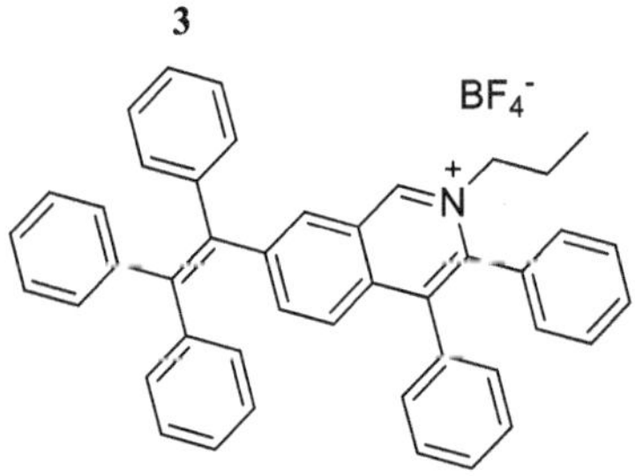 **TPE-IQ** **4**	Fluorescent agents for mitochondrial imaging and as apoptosis inducers, sensing of ROS, metal ions or pH change in mitochondria	44

high photostability and stable to microenvironment changes in cells. They claimed suitable modification led to synthesize efficient probes for different applications such as sensing of ROS, changes in pH and metal ions in mitochondria. Authors described that positively charged subunit in **2**, **3** and **4** facilitates the AIE probes toward mitochondria. Whereas the hydrophobic AIE core presume to display high tolerance toward the chaining microenvironment in cell line.

2.2 Biosensing

In biological field, luminescence technique is a perfect method as a direct visualization choice for small and biomacromolecules due to their superior sensitivity, high selectivity, low background noise, fast response and simplicity, etc. But the conventional chromophores exhibit ACQ effect which prohibited them to apply in biological system at high concentration.[21] Researchers established diverse sensing systems to detect and monitor important small biogenic such as amino acids,[45] biothiols,[46] amines,[47] adenosine triphosphate (ATP),[48] monosaccharides[49] and biomacromolecules such as proteins, polysaccharides, DNAs, lipids and enzymes too.[16,27] To overcome the limitations showed by ACQ, AIE material is an attractive choice due to their luminescence properties in the aggregate state, higher resistance to photobleaching, turn-probe and superior reliability. Such AIE chromophores brought technological innovation in biological applications too. In the field of life-sciences, medicinal, fermentation industry, pharmaceutical industry and biotechnological field the biogenic detection and monitoring is of crucial importance.

In this section we will compile some of the patents displaying AIEgens-based biosensing systems for small as well as biomacromolecules (Table 2).[50–52]

2.3 Cellular imaging, cell tracking, bacterial imaging

AIEgens are widely employed for imaging at the cellular and subcellular levels.[24,26] Moreover, the continuous cell tracking[53] using biocompatible contrast agents over a long period of time can offer valuable cellular processes such as monitoring tissue regeneration, understanding the genesis, development, invasion and metastasis of cancer information. Several biomedical imaging techniques and fluorescence emission imaging were demonstrated.[54] Among them, fluorescence imaging technique is most adapted labeling technique but it show some serious limitations. The conventional

Table 2 AIE based biosensing of small molecules and biomacromolecules.

Sr. No.	Representative example of AIEgens	Applications	References
1	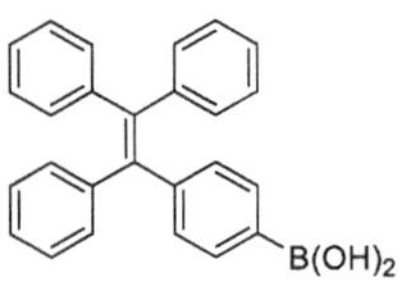 **TPEMB** **5** 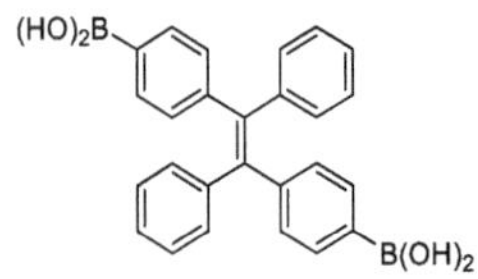 **TPEDB** **6**	Detection of D-glucose: Investigators claimed synthesis of water soluble tetraphenylethene chromophore having boronic acid functionalities and their AIE properties. They developed a method for pH detection in a sample solutions	50
2	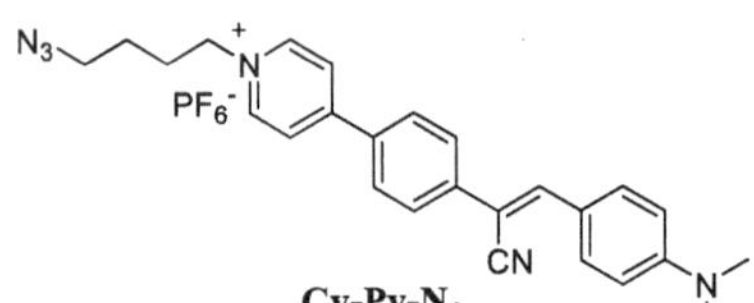 **Cy-Py-N_3** **7** 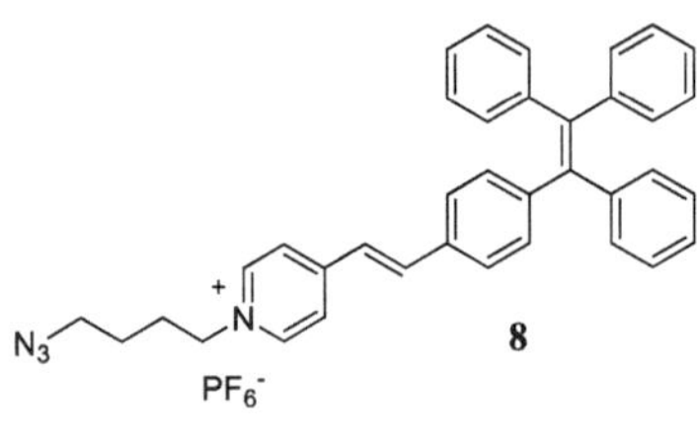 **8**	Tetraphenylethene based series of probes were synthesized by the investigator. They claimed the preparation method of AIE compounds and their utilization in detecting S-phase DNA and cell proliferation was also demonstrated. Investigators claim was upon binding with DNA, the fluorescence of molecule was turns-on. They also claimed as compare to Alexa-azide dye and a commercial DNA detection probe, the AIE chromophores described in this patent exhibit better photostability and sensitivity	51
3	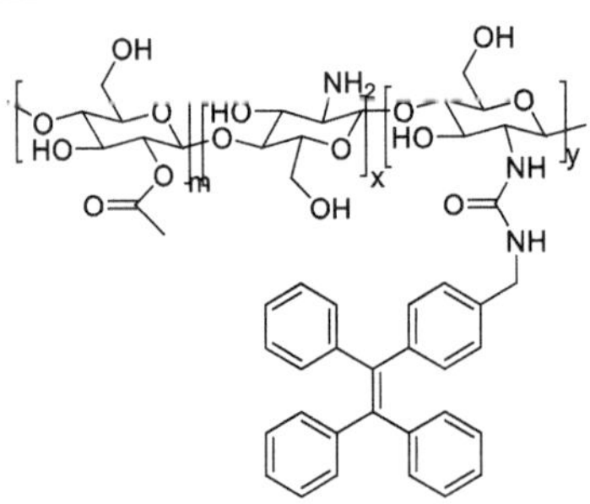 **9**	Authors claim, AIE fluorescent bioprobes comprised of TPE derivatives and chitosan, their utilization for labeling and detecting DNA and/or proteins as well as cell imaging and tracking	52

fluorescence probe display ACQ phenomenon, which in aggregate state shows fluorescence quenching. To overcome these shortcomings researchers developed AIE based emissive materials and some of them are patented as their intellectual property. The detail about structure of AIE probe and their applications are tabulated in Table 3.

Bacteria play key role in human life, some of them are useful in food industry, human health and environmental industry but others are pathogenic to affect human life. There are several fluorescence probes to monitor bacteria[58] and their activities but they display ACQ effect, which in turn are limited in use. To overcome these shortcoming and develop new efficient probe AIEgens are the natural choice. AIE probes in combination with non-specific AIE probes significantly improved their response toward Gram +Ve and Gram −Ve bacteria via electrostatic or hydrophobic interactions. Owing the importance of AIE materials and their utilization to direct visualization of bacteria and their type researchers developed new AIEgens and patented their work under an intellectual property right (Table 3).[55–57]

2.4 Image-guided therapy and visualization of treatment of cancer

Photothermal therapy (PTT) is one of the latest therapy method to cure the cancer.[59] PTT is a technique that convert light into heat energy to kill the cancer cells. PTT imaging technique employed to visualize the accurate location of cancer cells and monitor the real-time inflation of PTT agents in tumors. Recently AIEgens are employed in image-guided PTT for cancer treatment. AIEgens exhibit bright fluorescence in aggregated state and can enhance the imaging quality. AIEgens are the promising PTT agents with low toxicity and utilized for photoacoustic and photothermal imaging.[60]

Imaging technology in vivo is a technique employed for non-invasively visualizing and analyzing biological phenomenon occurring within a living body.[61] In recent decade, optical imaging using fluorescence and bioluminescence probe is superior to other techniques with respect to resolution and specificity. The utilization of such optical technology for visualization of deep tissue of cancer cells and treatment is difficult. To overcome this limitations, AIEgens are the emerging area of the research filed for vitalization of cancer treatment.

In this section of the chapter we highlighted the recent patent in AIEgens-based molecular architecture for imaging-guided PTT, PDT and visualization of treatment of cancer (Table 4).[62,63]

Table 3 Patented AIEgens for cell imaging, cell tacking and bacterial detection.

Sr. No.	Representative example of AIEgens	Applications	References
1	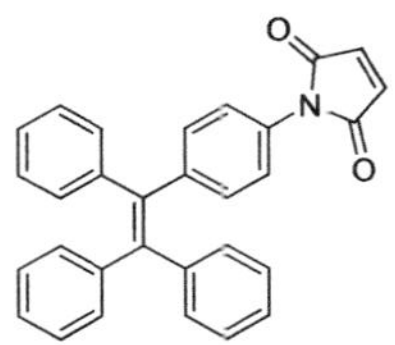 **10**	The TPE core conjugated with maleimide subunit are utilized for detecting thiol groups in biomolecules with high selectivity, sensitivity and is convenient and rapid technique	55
2	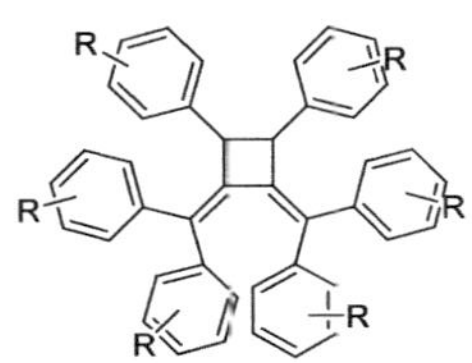 **11**	Luminogens with AIE characteristics can be used for long-term cell tracking. Such AIEgens can easily pass through cell membranes and track only the cytoplasm through high staining. Such AIEgens can be retained inside the cellular compartments and in aggregated state they are useful for long-term live cell imaging and tracking	56
3	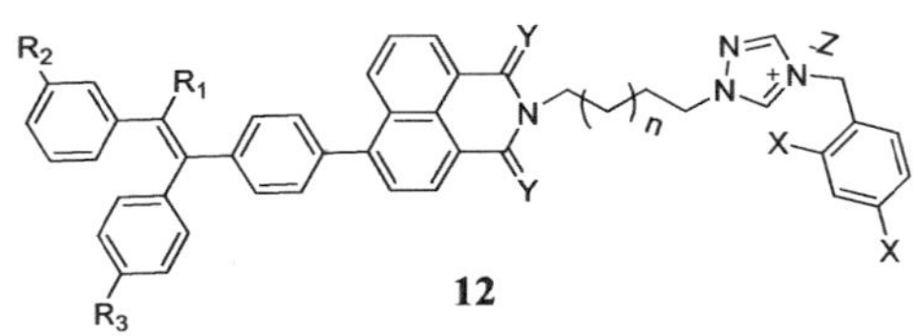 **12**	Luminogens containing both AIE and antibiotic compounds acts as an effective ROS generating AIEgens. This combined molecular architecture acts as antibacterial agents along with as a medicinal imaging tools to monitor the drug-bacteria interaction. Thus, the series of compounds patented in this patent display several functions such as imaging, monitoring, and antibacterial infection inhibition for treatment of patient in medical practice in hospitals	57

Table 4 Patented AIEgens for image-guided therapy, photodynamic therapy and visualization of treatment of cancer.

Sr. No.	Representative example of AIEgens	Applications	References
1	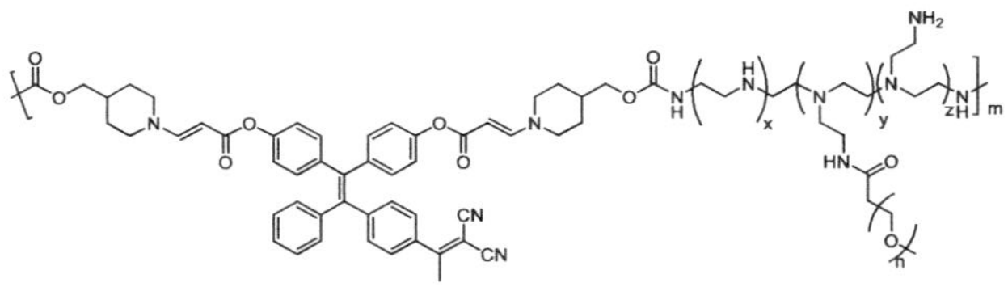 TPECM **13**	Oligomers and polymers with AIE properties are employed for drug tracking, delivery and identification and labeling of parts of cell or calls. These materials are utilized for bioimaging and image-guided photodynamic therapy	62
2	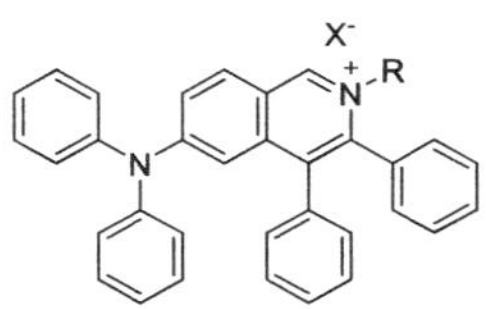 **14** 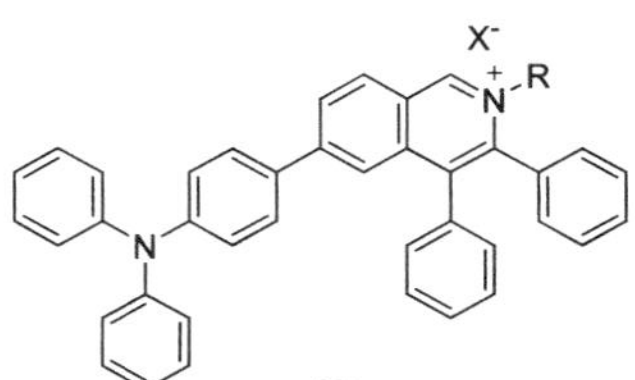 **15**	Authors claimed the intellectual property right on the utilization of AIE luminogens for visualization and treatment of cancer. Particularly they employed AIEgens for imaging and autophagy visualization. AIEgens with high emission properties as active theranostic agents for monitoring drug distribution. Such AIEgens utilized for visualization of anti-tumor activity to specific cancer cells, imaging and staining	63

2.5 AIE nanoparticles for intracellular imaging and protein carrier

In biomedical research optical bioimaging has become one of the prime technology in biomedical research. Such materials are utilized as a direct visualization tool for imaging biological and clinical processes. Fluorescent organic nanoparticles, carbon dots and semiconductor dots have been developed as advanced visualization technology.[64] At the same time they display some limitations such as ACQ effect at high concentration and the large organic molecular architecture display strong intermolecular π-π stacking interactions led in sever photobleaching. To overcome these limitations AIE based silica nanoparticles for intercellular imaging and protein carrier was developed and patented (Table 5).[65,66]

For intracellular imaging, the HeLa cells was treated with fluorescent visualizers such as fluorescent silica particles. AIE silica particles **17** upon interaction with HeLa cells exhibited the bright green emissions. The silica particles are entered into the HeLa cells through endocytosis and the nanoparticles are uptake by the cells. They claimed the synthesis of such nanoparticles bound to the biomolecules such as BSA and transported in cell lines.

The organic AIE materials such as **17** and **18** were utilized to synthesize nanoparticles through a modified nanoprecipitation method, using a mixture of DSPE-PEG_{2000} and DSPE-PEG_{5000}-Folate. The probes made up of compound **17** are treated to cancer cells in the area of tumor resulting into fluorescence emission at 1 and 3 h. The specific tumor targeting properties of **17** was also investigated on the same tumor-bearing mouse model. Herein much higher fluorescence was observed, suggesting probe **17** is an efficient chromophore for imaging in vivo.

2.6 Retardation of amyloid fibrillation of insulin

Recent year investigation on biomolecules indicating the understanding of pathogenesis of numerous diseases. It is well documented that fibrillation of amyloid proteins led to cause many neurodegenerative diseases, e.g., Parkinson's and Alzheimer's disease, Prion disease and Huntington's disease.[67–69] In diabetic patients at insulin injected site amyloid fibrillation found and are inductive of Parkinson's disease.[68,69] Therefore development of effective diagnostic tool for monitoring insulin aggregation and or amyloid fibril formation has been developed. The techniques such as transmission

Table 5 Patented AIEgens for intracellular imaging and protein carrier.

Sr. No.	Representative example of AIEgens	Applications	References
1	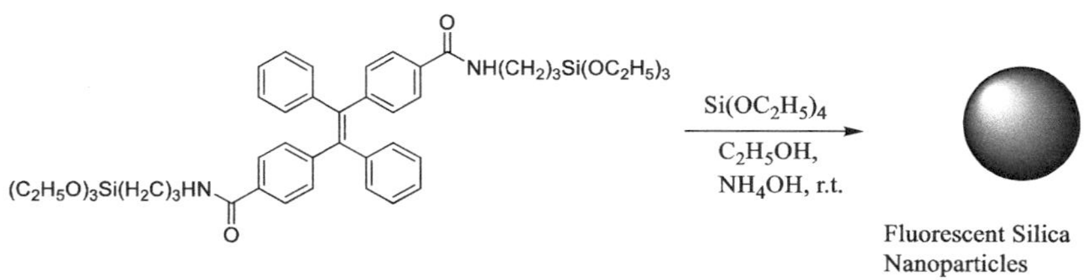 **16**	Fluorescent silica nanoparticles and fluorescent magnetic silica nanoparticles, exhibiting an AIE phenomenon their fabrication and use in intracellular imaging and a protein carrier was claimed by the authors in this patent	65

Continued

Table 5 Patented AIEgens for intracellular imaging and protein carrier.—cont'd

Sr. No.	Representative example of AIEgens	Applications	References
2	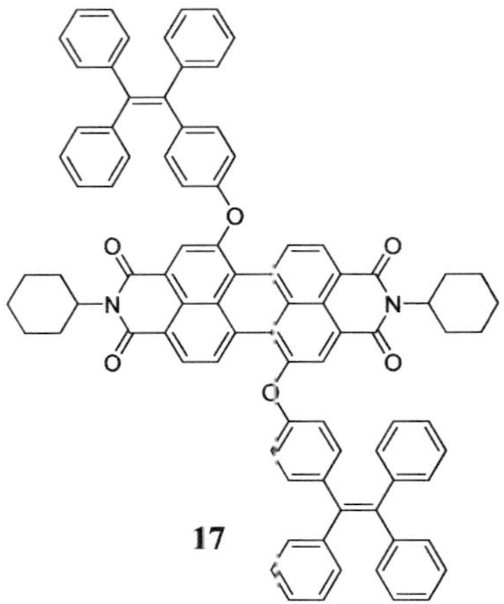 **17** 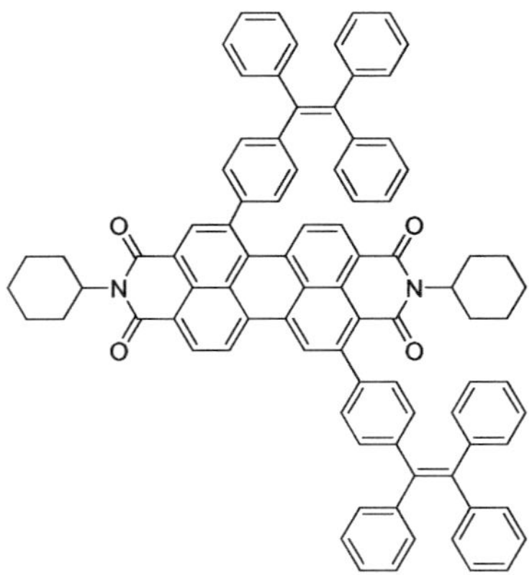 **18**	Fabrication of organic fluorescent chromophores and their applications for in vitro and in vivo bioimaging	66

electron microscopy (TEM), real-time light scattering, X-ray diffraction, atomic force microscopy (AFM), stopped-flow turbidimetry, fluorescence, circular dichroism (CD) and NMR spectroscopy were employed.[70] Among them fluorescence technique is the most widely and attractive technique used for intrinsic fluorescence of proteins.[16,27,71] Fluorescence method have been emerged as a mainstream area of research in science and technology, particularly in the field of biochemical and bioimaging fields. Such fluorescent chromophores are utilized for DNA sequencing, fluorescence activated cell sorting, high-through output screening and clinical diagnostics. In solution state these chromophores display high fluorescence properties but in aggregated state or in high concentration they exhibited ACQ effect. Last two decades are witnessed the rise of novel fluorescence materials with AIE properties. AIE chromophores yields high fluorescence emission properties in its aggregate formation.

1,1,2,2-tetrakis[4-(2-triethylammonioecthoxy)-phenyl] ethnetetrabromide (TTAPE) **19** (Table 6) is non-emissive in dilute solution but becomes emissive in its concentrated solution as well as in aggregation state. TTAPE bound to DNA via electrostatic interaction and the emission is turned on. The binding process is reversible in presence of a competitive cation. TTAPE acts as sensitive probe for visualization of DNA via binding

Table 6 Patented AIEgens for retardation of amyloid fibrillation of insulin.

Sr. No.	Representative example of AIEgens	Applications	References
1	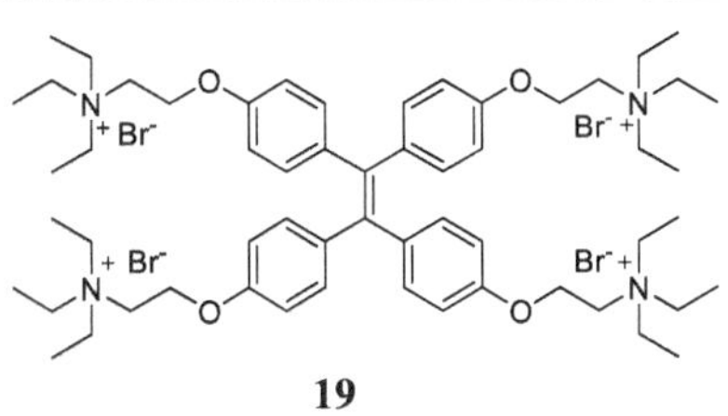 **19** 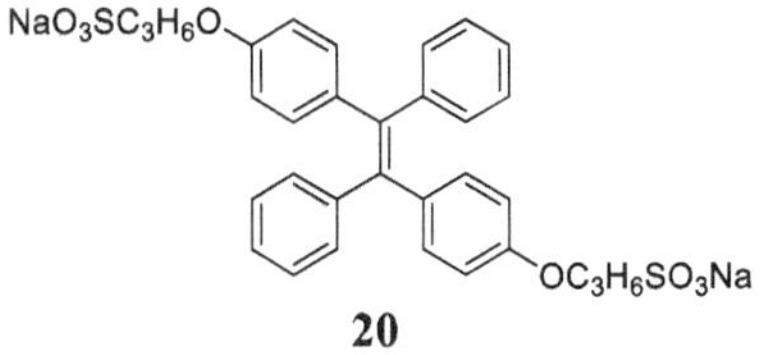 **20**	In this patent, investigator described preparation of water soluble conjugated polyene AIEgens. These polyenes were employed for detection of biomacromolecules, which are involved in monitoring and retarding amyloid protein fibril formation in vitro and in vivo as well as in developing anti-cancer drugs	72

to a secondary structure of G-quadruplex. The AIE characteristic of TTAPE enables to monitor folding process of G1 in aqueous media.

Sodium 1,2-diphenyl-1,2-bis(4,4′-(3-sulfonato)propoxyl)phenylethylene (TPE-SO_3) **20** (Table 6) was employed for detection of amyloid fibril formation. The solution of **20** becomes luminescent with an increase in concentration of insulin fibrils. The increase in emission of **20** is useful to monitor a linear increase of insulin fibril fraction. The TPE-SO_3 display a retardation effect on amyloid fibrillation. Authors claimed their investigation how exponential decrease in growth rate of insulin fibrillation against the concentration of TPE-SO_3.

2.7 Photoactivatable probe

In biochemical and biological research, photoactivatable probes with the change in fluorescent state from non-emissive to emissive, upon irradiation at the particular wavelengths have been an attractive tool. Photoactivatable fluoroprobes should possess an important properties such as easy synthesis, efficient and rapid photoactivation, and high photostability, specificity, less harmful and perturbative to biological system and environmentally insensitive. The common photoactivatable fluoroprobes are fluorescent proteins,[73–75] though they are genetically targeted but display some drawbacks, i.e., less brightness, easy photobleaching, stimuli responsive toward change in pH than small organic molecules. Therefore, small organic molecules[76] possessing an important properties such as brighter emission, less perturbation and more readily structure manipulation have attracted researcher's attention. The reported small organic molecules are mainly caged-fluorescein, caged-rhodamine, caged-coumarin, caged-BODIPY, azido-DCDHF pull-push probe and intramolecular tetrazole-alkene.[77] However, these chromophores still showed some limitations. To overcome these drawbacks, in recent year researcher developed AIE based fluoroprobes for the bioresearch field advancement. In the following patent, investigators reported AIE based photoactivatable fluoroprobes.

The prepared photoactivatable probe o-TPP3M (**21**) (Table 7) can be photoactivated by using visible or UV-light and are mitochondria specific. o-TPP3M probe is used in spatiotemporal selective cells, i.e., HeLa cells, organelles and macromolecules imaging. o-TPP3M undergoes photo-oxidative dehydrocyclization to lead photo-irradiated state with strong emission.

Table 7 Patented AIE photoactivatable probe.

Sr. No.	Representative example of AIEgens	Applications	References
1	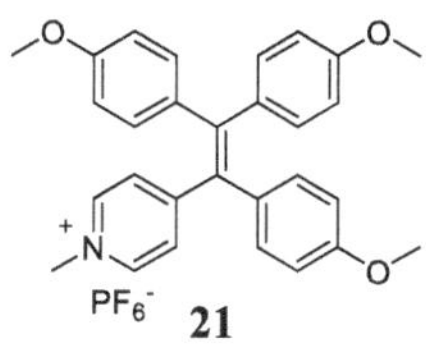 **21** o-TPP3M	Photoactivatable fluoroprobe-TPP3M is photoactivated to a strong emissive c-TPP3M. The high selectivity of o-TPP3M in HeLa living cells indicate that this probe was very good entity in bioresearch with highly spatiotemporal resolution. o-TPP3M display no toxicity and is specific to mitochondria without any decoration. This molecular architecture provide a new platform for fabrication of novel photoactivatable fluorophores in bioresearch	78

2.8 Method of detecting human serum albumin in biological fluids

Early diagnosis of a chronic disease such as renal damage can facilitate successful treatment. Urine is an ideal fluid to monitor and diagnosis of renal damage. Human serum albumin (HAS), major protein in blood plasma,79 is generally not present in urine. Kidney malfunctions and glomerular permeability lead to have HAS in urine.[80] Therefore, detection of presence of HAS and monitoring its concentration in urine is an important task for researchers.[81] The common HAS detection methods utilized are sulphosalicylic acid test, Heller's test, colorimetric method, i.e., Bradford and Lowry assay, microchip electrophoresis, capillary electrophoresis, high performance liquid chromatography, enzymatic immunoassay or turbidimetric inhibitory immunoassay, dry chemistry test paper, colloidal gold immune stripand fluorescence method.[82] These probes display twisted intramolecular charge transfer (TICT) mechanism. Such small organic molecules display drawbacks such as water-insoluble, organic co-solvent is required and responsive to polarity of solvent and changes in pH of a urine sample. Therefore, highly efficient HSA monitoring and detection method in a biological fluid is highly desirable. To overcome these limitations, novel material based on AIE properties of the molecular architecture are an attractive choice. Such fluorescent probe is non-emissive in absence of HAS. When HAS is present in biological fluid such as urine, the binding of AIE probe with HAS yields high emission.

Table 8 Patented AIE probe for HAS detection.

Sr. No.	Representative example of AIEgens	Applications	References
1	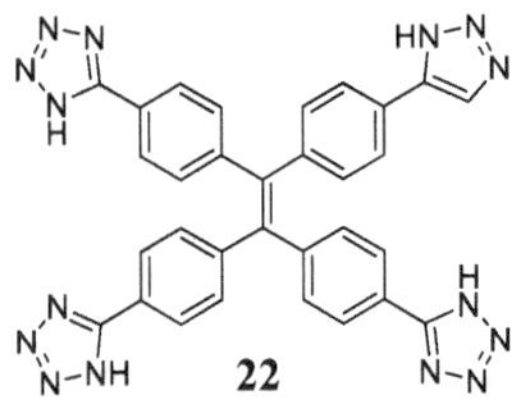 22 **TPE-4TA** 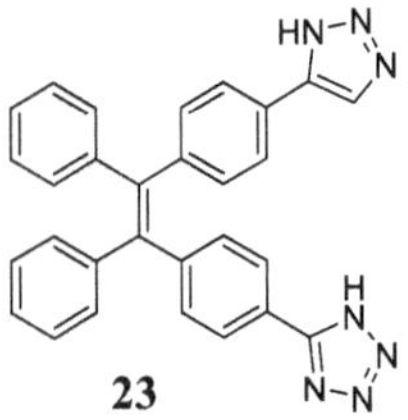 23 **TPE-2TA-a**	AIE molecular architecture was employed for detection of HAS by irradiating the mixture with ultraviolet light. The probe yields high fluorescence upon binding with HAS in biological fluid, i.e., urine. The prepared compound is water soluble. This technique can be utilized in hospitals and if the device is portable then can be used by a patient at home	83

The probes **22** (Table 8) and **23** (Table 8) are non-emissive in PBS buffer at neutral pH (physiological pH 7.4). With the addition of HAS into the PBS buffer solution, these chromophores bind with HAS and the resulting complex can show an intense emission properties. These probes were utilized for detection of HAS in biological fluid such as urine. The method is also impressive for quantitative determination of HAS in urine.

3. Future prospectus

The investigation of molecular architectures with AIE phenomenon has reaching to practical implications. AIE chromophores are promising candidates with turn-on characteristics are useful for onsite-screening, field-trial and real-world applications. The higher fluorescence emission efficiencies of AIEgens in aggregated and solid state are beneficial compared to ACQ materials. Recent year's publication and patent are witnessing the fascinating applications of AIEgens in the field of biological and medical sciences and engineering. Further study of AIE is bring more astonishing results with innovative applications in the field of biomedical sciences as well as materials sciences.

Acknowledgments

S.V.B. (IICT) is grateful for financial support from The DAE-BRNS under the project No. 58/14/01/2020-BRNS/37047 and the Director, CSIR-IICT, Hyderabad, IICT/Pubs./2020/316. Dedicated to Prof. Wamanrao N. Jadhav, Ex-dean, Swami Ramanand Teerth Marathwada University, Nanded, India on his superannuation.

References

1. Newton HE. Luminescent organisms. *Am Sci.* 1952;40:468–481.
2. Braslavsky SE. Glossary of terms used in photochemistry, 3rd edition, (IUPAC recommendations 2006). *Pure Appl Chem.* 2007;79:293–465.
3. Valeur B, Berberan-Santos MN. A brief history of fluorescence and phosphorescence before the emergence of quantum theory. *J Chem Educ.* 2011;88:731–738.
4. Monardes N. *Historia medicinal de lascosasque se traen de nuestras Indians occidentalesquesirven en medicina.* Padilla Libros; 1565:1.
5. Acuña AU. More thoughts on the Narra tree fluorescence. *J Chem Educ.* 2007;84:231.
6. de Sahagún. In: De Bustamante CM, ed. *Florentine Codex, Historia General de lascosas de Neva España.* vol III; 1577. p. 95 and p. 231.
7. Acuña AU, Amat-Guerri F, Morcillo P, Liras M, Rodriguez B. Structure and formation of the fluorescent compound of lignum nephriticum. *Org Lett.* 2009;11:3020–3023.
8. Brewster D. XIX. On the colors of natural bodies. *Trans R Soc Edinburgh.* 1834;12:538.
9. Brewster D. On the colours of natural bodies. *London Edinburgh Philos Mag J Sci.* 1836; VIII:468.
10. Hongsbaum M, Willcox M. In: Willcox M, Bodeker G, Rasoanaivo P, eds. *Chinchona in Traditional Medicinal Plants and Malaria.* Boca Raton, Florida: USA: CRC Press; 2005:22. Chapter 2.
11. Herschel JFW. On a case of superficial colours presented by a homogenous liquid internally colourless. *Philos Trans R Soc London.* 1845;135:143–145.
12. Stokes GG. On the change of refrangibility of light. *Philos Trans R Soc London.* 1852;142:463.
13. Duval R, Duplais C. Fluorescent natural products as probes and tracers in biology. *Nat Prod Rep.* 2017;34:161–193.
14. Jabłoński A. Efficiency of anti-stokes fluorescence in dyes. *Nature.* 1933;131:839–840.
15. Jaffe HH, Miller AL. The fates of electronic excitation energy. *J Chem Educ.* 1966;43:469.
16. Wu D, Sedgwick AC, Gunnlaugsson T, Akkaya EU, Yoon J, James TD. Fluorescent chemosensors: the past, present and future. *Chem Soc Rev.* 2017;46:7105–7123.
17. Brennan JD. In: Meyers RA, ed. *Encyclopedia of Analytical Chemistry.* John Wiley & Sons, Ltd; 2016:1–52.
18. Sánchez-Vergara ME, Hamui L, González HS. New approaches in flexible organic field-effect transistors (FETs) using InClPc. *Materials.* 2019;12:1712.
19. Im Y, Kim JH, Lee DR, Oh CS, Yook KS, Lee JY. Recent progress in high-efficiency blue-light-emitting materials for organic light-emitting diodes. *Adv Funct Mater.* 2017;27:1603007.
20. Fraiji LK, Hayes DM, Werner TC. Static and dynamic fluorescence quenching experiments for the physical chemistry laboratory. *J Chem Educ.* 1992;69:424–428.
21. (a) Birks JB. *Photophysics of Aromatic Molecules.* London, U.K: Wiley; 1970. (b) Zhelev Z, Ohba H, Bakalova R. Single quantum dot-micelles coated with silica shell as potentially non-cytotoxic fluorescent cell tracers. *J Am Chem Soc.* 2006;128:6324–6325.
22. Würthner F. Aggregation-induced emission (AIE): a historical perspective. *Angew Chem Int Ed.* 2020;59:2–7.

23. Luo J, Xie Z, Lam JWY, et al. Aggregation-induced emission of 1-methyl-1, 2,3,4,5-pentaphenylsilole. *Chem Commun.* 2001;1740–1741.
24. Mei J, Leung NLC, Kwok RTK, Lam JWY, Tang BZ. Aggregation-induced emission: together we shine, united we soar. *Chem Rev.* 2015;115:11718–11940.
25. Mei J, Hong YN, Lam JWY, Qin AJ, Tang YH, Tang BZ. Aggregation-induced emission: the whole is more brilliant than the parts. *Adv Mater.* 2014;26:5429–5479.
26. Wang Y, Nie J, Fang W, et al. Sugar-based aggregation-induced emission luminogens: design, structures, and applications. *Chem Rev.* 2020;120:4534–4577. and references therein.
27. Berezin MY, Achilefu S. Fluorescence lifetime measurements and biological imaging. *Chem Rev.* 2010;110:2641–2684. and references therein.
28. Specht EA, Braselmann E, Palmer AE. A critical and comparative review of fluorescent tools for live-cell imaging. *Annu Rev Physiol.* 2017;79:93–117.
29. Liang GD, Lam JWY, Qin W, Li J, Xie N, Tang BZ. Molecular luminogens based on restriction of intramolecular motions through host-guest inclusion for cell imaging. *Chem Commun.* 2014;50:1725–1727.
30. Hu R, Gómez-Durán CFA, Lam JWY, et al. Synthesis, solvatochromism, aggregation-induced emission and cell imaging of tetraphenylethene-containing BODIPY derivatives with large stokes shifts. *Chem Commun.* 2012;48:10099–10101.
31. Li Y, Wu Y, Chang J, Chen M, Liu R, Li F. A bioprobes based on aggregation induced emission (AIE) for cell membrane tracking. *Chem Commun.* 2013;49:1133–11337.
32. Yu Y, Feng C, Hong Y, et al. Cytophilic fluorescent bioprobes for long-term cell tacking. *Adv Mater.* 2011;23:3298–3302.
33. Ow Y-LP, Green DR, Hao Z, Mak TW. Cytochrome *c*: functions beyond respiration. *Nat Rev Mol Cell Biol.* 2008;9:532–542.
34. Hoye AT, Davoren JE, Wipf P, Fink MP, Kagan VE. Targeting mitochondria. *Acc Chem Res.* 2008;41:87–97.
35. Gardner A, Boles RG. Is a "mitochondrial psychiatry" in the future? A review. *Curr Psychiatry Rev.* 2005;1:255–271.
36. Lesnefsky EJ, Moghaddas S, Tandler B, Kerner B, Hoppel CL. Mitochondrial dysfunction in cardiac disease: ischemia-reperfusion, aging, and heart failure. *J Mol Cell Cardiol.* 2001;33:1065–1089.
37. Dickinson BC, Srikun D, Chang CJ. Mitochondrial-targeted fluorescent probes for reactive oxygen species. *Curr Opin Chem Biol.* 2010;14:50–56.
38. Dickinson BC, Chang CJ. A targetable fluorescent probe for imaging hydrogen peroxide in the mitochondria of living cells. *J Am Chem Soc.* 2008;130:9638–9639.
39. Masanta G, Heo CH, Lim CS, Bae SK, Cho BR, Kim HM. A mitochondria-localized two-photon fluorescent probe for ratiometric imaging of hydrogen peroxide in live tissue. *Chem Commun.* 2012;48:3518–3520.
40. Forman HJ, Zhang H, Rinna A. Glutathione: overview of its protective roles, measurement, and biosynthesis. *Mol Aspects Med.* 2009;30:1–12.
41. Zhang H, Wang CX, Wang K, Xuan XP, Lv QZ, Jiang K. Ultrasensitive fluorescent ratio imaging probe for the detection of glutathione ultratrace change in mitochondria of cancer cells. *Biosens Bioelectron.* 2016;85:96–102.
42. Lim SY, Hong KH, Kim DI, Kwon H, Kim HJ. Tunable heptamethine–azo dye conjugate as an NIR fluorescent probe for the selective detection of mitochondrial glutathione over cysteine and homocysteine. *J Am Chem Soc.* 2014;136:7018–7025.
43. Neto BAD, Carvalho PHPR, Santos DCBD, et al. Synthesis, properties and highly selective mitochondria staining with novel, stable and superior benzothiadiazole fluorescent probes. *RSC Adv.* 2012;2:1524–1532.

44. Tang BZ, Hong Y, Chen S, Leung WT, Zhao E, Deng H. *Photostable AIE Luminogens for Specific Mitochondrial Imaging and its Method of Manufacturing Thereof*; 2016. US9,315,465 B2.
45. Zhou Y, Yoon J. Recent progress in fluorescent and colorimetric chemosensors for detection of amino acids. *Chem Soc Rev*. 2012;41:52–67.
46. (a) Riederer BM. Oxidation proteomics: the role of thiol modifications. *Curr Proteomics*. 2009;6:51–62. (b) Zhang M, Yu M, Li F, et al. A highly selective fluorescence turn-on sensor for cysteine/homocysteine and its application in bioimaging. *J Am Chem Soc*. 2007;129:10322–10323.
47. (a) Klockow JL, Hettie S, Glass TE. Fluorescent sensors for amines. In: Atwood JL, ed. *Comprehensive Supramolecular Chemistry II*. Elsevier; 2017 [Ed.-in-Chief]. (b) Kannan SK, Sudalaimani S, Pandiaraj M, Giribabu K, Kathiresan M. A review on chemical and electrochemical methodologies for the sensing of biogenic amines. *Anal Methods*. 2020;12:3438–3453.
48. Jun YW, Sarkar S, Kim KH, Ahn KH. Molecular probes for fluorescence imaging of ATP in cells and tissues. *ChemPhotoChem*. 2019;3:214–219.
49. (a) Jovanovic L. The importance of postprandial glucose concentration: lessons learned from diabetes and pregnancy. *Drug Dev Res*. 2006;67:591–594. (b) Saltiel AR, Kahn CR. Insulin signalling and the regulation of glucose and lipid metabolism. *Nature*. 2001;414:799–806. (c) James TD, Sandanayake KRAS, Shinkai S. Saccharide sensing with molecular receptors based on boronic acid. *Angew Chem Int Ed Engl*. 1996; 35:1910–1922.
50. Tang BZ, Liu Y, Sun J, Qin A, Hu R. *Specific Detection of D-Glucose by a Tetraphenylethene-Base Fluorescent Sensor*; 2012. US 2012/0283703 A1.
51. Tang BZ, Zhao Y, Guo Z. *Photostable AIE Fluorogens for Accurate and Sensitive Detection of S-Phase DNA Synthesis and Cell Proliferation*; 2016. WO 2016/165616 A1.
52. Tang BZ, Ng KM, Luo Q, et al. *Aggregation Induced Emission of Fluorescent Bioprobes and Methods of Using the Same*; 2013. US9,618,453 B2.
53. (a) Goodman LR. The beatles, the Nobel prize, and CT scanning of the chest. *Radiol Clin North Am*. 2010;48:1–7. (b) Hounsfield GN. Computerized transverse axial scanning (tomography). 1. Description of system. *Br J Radiol*. 1973;46:1016–1022.
54. Ni J-S, Li Y, Yue W, Liu B, Li K. Nanoparticle-based cell trackers for biomedical applications. *Theranostics*. 2020;10:1923–1947. and references therein.
55. Tang BZ, Liu Y, Lam WY, Yu Y. *Luminogen Compounds and the Use of the Same for Biosensing and Cellular Imaging*; 2012. US9,475,768B2.
56. Tang BZ, Ng KM, Luo Q, et al. *Aggregation Induced Emission Active Cytophilic Fluorescent Bioprobes for Long-Term Cell Tracking*; 2013. US9,409,928 B2.
57. Tang BZ, Li Y, Zhao Z. *A Bifunctional Aggregation-Induced Emission Luminogen for Monitoring and Killing of Multidrug-Resistant Bacteria*; 2020. WO2020/011228A1.
58. (a) Miyanaga L, Takano S, Morono Y, Hori K, Unno H, Tanji Y. Optimization of distinction between viable and dead cells by fluorescent staining and its application to bacterials consortia. *Biochem Eng J*. 2007;37:56–61. (b) Cheng MS, Lau SH, Chow VT, Toh C-S. Membrane-based electrochemical nanobiosensor for *Escherichia coli* detection and analysis of cell viability. *Environ Sci Technol*. 2011;45:6453–6459.
59. (a) Cheng L, Wang C, Feng LZ, Yang K, Liu Z. Functional nanomaterials for phototherapies of cancer. *Chem Rev*. 2014;114:10869–10939. (b) Jung HS, Verwilst P, Sharma A, Shin J, Sessler JL, Kim JS. Organic molecule-based photothermal agents: an expanding photothermal therapy universe. *Chem Soc Rev*. 2018;47:2280–2297. (c) Liu YJ, Bhattarai P, Dai ZF, Chen XY. Photothermal therapy and photoacoustic imaging via nanotheranostics in fighting cancer. *Chem Soc Rev*. 2019;8:2053–2108.

60. Wang J, Li J, Wang L, Han T, Wang D, Tang BZ. AIEgen-based polymer nanocomposites for imaging-guided photothermal therapy. *ACS Appl Polym Mater.* 2020;2:4306–4318.
61. Imamura T, Saiou T, Kawakami R. In vivo optical imaging of cancer cell function and tumor microenvironment. *Cancer Sci.* 2018;109:912–918.
62. Liu B, Yuan Y, Feng G, et al. *Polymers and Oligomers With Aggregation-Induced Emission Characteristic for Imaging and Image-Guided Therapy*; 2017. US2017/0168041A1.
63. Tang BZ, Zhao E, Gui C, et al. *AIE Luminogens for Visualization and Treatment of Cancer*; 2016. WO2016/206615A1.
64. Wang EC, Wang AZ. Nanoparticles and their applications in cell and molecular biology. *Integr Biol (Camb).* 2014;6:9–26. and references therein.
65. Tang BZ, Lam WY, Liu J, Mahtab F, Liu Y, Yu Y. *Silica Nanoparticles With Aggregation Induced Emission Characteristic as Fluorescent Bioprobes for Intracellular Imaging and Protein Carrier*; 2013. US 9, 518, 921 B2.
66. Tang BZ, Qin W, Liu J, et al. *Biocompatible Nanoparticles With Aggregation Induced Emission Characteristic as Fluorescent Bioprobes and Methods of Using the Same for In Vitro and In Vivo Imaging*; 2014. US2014/0328764 A1.
67. Lednev IK. Amyloid fibrils: the eighth wonder of the world in protein folding and aggregation. *Biophys J.* 2014;106:1433–1435.
68. Choi ML, Gandhi S. Crucial role of protein oligomerization in the pathogenesis of Alzheimer's and Parkinson's diseases. *FEBS J.* 2018;285:3631–3644.
69. Chiti F, Dobson CM. Protein misfolding, functional amyloid, and human disease. *Annu Rev Biochem.* 2006;75:333–366.
70. (a) Rubin N, Perugia E, Goldschmidt M, Fridkin M, Addadi L. Chirality of amyloid suprastructures. *J Am Chem Soc.* 2008;130:4602–4603. (b) Nettleton EJ, Tito P, Sunde M, Bouchard M, Dobson CM, Robinson CV. Characterization of the oligomeric states of insulin in self-assembly and amyloid fibril formation by mass spectrometry. *Biophys J.* 2000;79:1053–1065. (c) Manno M, Craparo EF, Podestà A, et al. Kinetics of different processes in human insulin amyloid formation. *J Mol Biol.* 2007;366: 258–274. (d) Domike KR, Donald AM. Thermal dependence of thermally induced protein spherulite formation and growth: kinetics of β-lactoglobulin and insulin. *Biomacromolecules.* 2007;8:3930–3937. e Pratim Bose P, Chatterjee U, Xie L, Johansson J, Göthelid E, Arvidsson PI. Effects of Congo red on $A\beta_{1-40}$ fibril formation process and morphology. *ACS Chem Nerosci.* 2010;1:315–324. f Levine H. Thioflavine T interaction with synthetic Alzheimer's disease beta-amyloid peptides: detection of amyloid aggregation in solution. *Protein Sci.* 1993;2:404–410. g Groenning M, Norrman M, Flink JM, et al. Binding mode of thioflavin T in insulin amyloid fibrils. *Struct Biol.* 2007;159:483–497. h Rodriguez-Rodriguez C, Rimola A, Rodriguez-Santiago L, et al. Crystal structure of thioflavin-T and its binding to amyloid fibrils: insights at the molecular level. *Chem Commun.* 2010;46:1156–1172.
71. Bekard IB, Dunstan DE. Tyrosine autofluorescence as a measure of bovine insulin fibrillation. *Biophys J.* 2009;97:2521–2531.
72. Tang BZ, Hong Y, Chen S, Kwok RTK. *Water Soluble AIE Luminogen for Monitoring and Retardation of Amyloid Fibrillation of Insulin*; 2012. US2012/0172296 A.
73. Lukyanov KA, Chudakov DM, Lukyanov S, Verkhusha VV. Photoactivatable fluorescent proteins. *Nat Rev Mol Cell Biol.* 2005;6:885–890.
74. Shaner NC, Patterson GH, Davidson MW. Advances in fluorescent protein technology. *J Cell Sci.* 2007;120:4247–4260.
75. Patterson GH, Lippincott-Schwartz J. A photoactivatable GFP for selective photolabeling of proteins and cells. *Science.* 2002;297:1873–1877.
76. Maurel D, Banala S, Laroche T, Johnsson K. Photoactivatable and photoconvertible fluorescent probes for protein labelling. *ACS Chem Biol.* 2010;5:507–516.

77. Li WH, Zheng G. Photoactivatable fluorophores and techniques for biological imaging applications. *Photochem Photobiol Sci*. 2012;11:460–471.
78. Tang BZ, Gu X, Zhao E, Lin R. *Photoactivatable Bioprobes: Design, Method of Preparation and Applications*; 2016. WO2016/070854A1.
79. Peters T. All about albumin: biochemistry. In: Peters T, ed. *Genetics and Medical Applications*. San Diego, CA: Academic Press Limited; 1996:9–54.
80. Pathology Harmony, UK. Harmonisation of reference intervals, pathology harmony group. In: *Clinical Biochem. Outcomes*. Pathology Harmony Group; 2011. January.
81. Speeckaert MM, Speeckaert R, van de Voorde L, Delanghe JR. Immunochemically unreactive albumin in urine: fiction or reality? *Crit Rev Clin Lab Sci*. 2011;48:87–96.
82. (a) Fanali G, Masi AD, Trezza V, Marino M, Fasano M, Ascenzi P. Human serum albumin: from bench to bedside. *Mol Aspects Med*. 2012;33:209–290. (b) Chen M, Xiang X, Wu K, He H, Chen H, Ma C. A novel detection method of human serum albumin based on the poly(thymine)-templated copper nanoparticles. *Sensors*. 2017;17:2684. and references therein.
83. Tang BZ, Xie S, Tu Y. *Method for Detecting Human Serum Albumin in Biological Fluids*; 2020. WO2020/043087 A1.

Index

Note: Page numbers followed by "*f*" indicate figures, "*t*" indicate tables, and "*s*" indicate schemes.

B

C

Q

U

V

Z

FSC
www.fsc.org
MIX
Papier | Fördert
gute Waldnutzung
FSC® C083411

Kolibri 360 GmbH
Ferdinand-Jühlke-Straße 7
99095 Erfurt, Deutschland
produktsicherheit@kolibri360.de